Contents

Maths Skills

for A Level

Business Studies

Diane Mansell

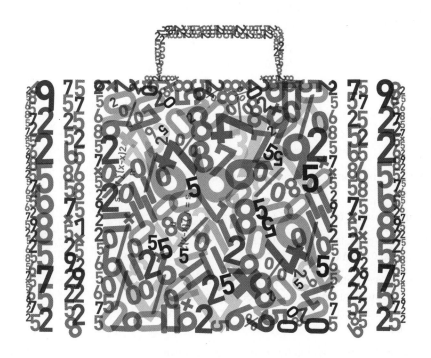

OXFORD

Windsor and Maidenhead

Great Clarendon Street, Oxford, OX2 6DP, United Kingdom

Oxford University Press is a department of the University of Oxford.
It furthers the University's objective of excellence in research, scholarship,
and education by publishing worldwide. Oxford is a registered trade mark of
Oxford University Press in the UK and in certain other countries

British Library Cataloguing in Publication Data
Data available

978-1-4085-2707-8

1 3 5 7 9 10 8 6 4 2

Printed in China

Acknowledgements
Thanks are due to Peter Stimpson for his contribution in the development of this
book.

Page make-up and illustrations: Tech-Set Ltd, Gateshead

Table on 'Discount Factors @ 1% to 10% pa' sourced from The Department of
Finance and Personnel. © Crown copyright 2010.

How to use this book

This workbook has been written to support the development of key mathematics skills required to achieve success in your A Level Business Studies course. It has been devised and written by teachers and the practice questions included reflect the **exam-tested content** for AQA, OCR and Cambridge specifications.

The workbook is structured into sections with each section having a clear business topic. Then, each spread covers a mathematical skill or skills that you may need to practise. Each spread offers the following features:

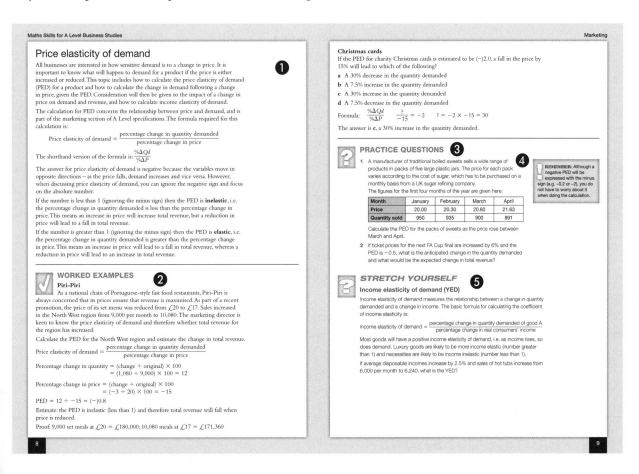

❶ *Main text* outlines the mathematical skill or skills covered within the spread.

❷ *Worked example* – each spread will have one or two worked examples to demonstrate calculations and working-out methods.

❸ *Practice questions* are to increase your confidence using contextual examples. All answers are available.

❹ *Remember* is a useful box that will offer you tips, hints and other snippets of useful information.

❺ *Stretch yourself* – some of the spreads may also contain a few more difficult questions at the end to stretch your mathematical knowledge and understanding.

Market size, share and growth

Most businesses will collect a range of data to assess the market they are in or wish to operate in. The formulae that may be required are for market size, market share and market growth. These are given below.

Note that the market share and market growth calculations are expressed as percentages.

Market size

The size of the market is measured by adding the total sales of all businesses in the market. It is expressed in terms of either the volume of sales in units or the value of sales in currency.

Market size by volume = volume of sales in the market
(e.g. 2 million car sales in the UK in 2013)

Market size by value = total volume of sales × average market price

or

$$\text{Market size by value} = \frac{\text{business sales}}{\text{\% market share}} \times 100$$

> **REMEMBER:** Market size by value can also be calculated as total sum of sales value of all businesses in the market.

WORKED EXAMPLES

Smartphones

Analysts suggest that the volume of sales in the global market for internet-enabled smartphones is about 225 million. The average price of a phone has fallen rapidly to around £230.

Calculate the value of the global smartphone market.

Value of sales in the market = total volume of sales × average market price
= 225,000,000 × £230 = £51,750,000,000

Special chilli jam

Sarah is really pleased with her sales figures for last year. She achieved 5% of the chilli jam market in the UK by the end of her fourth year of trading. Her sales revenue was also very healthy at £750,000.

Calculate the value of the chilli jam market in the UK.

Market size by value = (business sales ÷ % market share) × 100
= (£750,000 ÷ 5) × 100 = £15,000,000

PRACTICE QUESTIONS

1 The hotels and restaurants and shops on the island of Jersey buy on average 10,000 tomatoes per year at an average price of £2.50 per pack of five.
Calculate the value of the tomato market on Jersey.

2 ABC Taxis operates in and around Derby in the East Midlands. It runs a small fleet of cars and minibuses and believes that it controls 35% of the local market, including travel to the airport at Castle Donnington. The average fare per journey is £15 and ABC completes 25,000 journeys per year.
Calculate the value of the taxi market in the Derby area to the nearest pound.

Market share

This is the proportion of a total market accounted for by one product or company. It is expressed as a percentage of the total value of sales in the market.

$$\text{Market share} = \frac{\text{value of firm A's sales}}{\text{value of total market sales}} \times 100$$

WORKED EXAMPLE

Fiona's flower power

There are eight florists in Market Harwell and between them they have an annual turnover of £175,000. Fiona Parker believes that she now has a better location, having moved close to the hospital and the railway station. Since the move, her sales revenue has increased to £35,000 per year and she hopes that she has also increased her share of the market. In her previous shop, Fiona's market share was 15%. Calculate her new share of the market.

Market share = (value of firm A's sales ÷ value of total market sales) × 100

= (35,000 ÷ 175,000) × 100 = 20%

PRACTICE QUESTION

3 John Middleton grows tomatoes on Jersey and supplies local businesses as well as shipping fresh tomatoes to the UK mainland. His revenue from sales on Jersey is £15,000 per year. Calculate his share of the Jersey tomato market.
Note that you will need your answer from practice question 1 to complete this answer.

> **REMEMBER:** Market share and market growth calculations are expressed as percentages.

Market growth

This is the measurement of the change in the size of a market. It is usually expressed as a percentage change from the original market size by value.

$$\text{Market growth} = \frac{\text{change in size of market by value}}{\text{original size of market by value}} \times 100$$

> **REMEMBER:** Growth of a market does not occur when new firms enter the market. This just increases the competition within the market.

WORKED EXAMPLE

Online dating

Love For Life is a UK online dating site, concentrating on the young graduate market. The business has grown rapidly over the past five years from 5,000 members in the first year of business to 40,000 at the end of the last financial year.

Calculate the percentage change in membership from year 1 to the end of the last financial year.

Market growth = (change in size of market by value ÷ original size of market by value) × 100

Change in membership = 40,000 − 5,000 = 35,000

Original membership = 5,000

Market growth = (35,000 ÷ 5,000) × 100 = 700%

PRACTICE QUESTION

4 The *Nottingham Telegraph* is a local newspaper planning to launch an online version next year to take account of the change in readers' buying habits. The marketing department has made the following projections about online readership over the next five years:

Year	1	2	3	4	5
Readership	25,000	32,000	37,000	43,000	45,000

Calculate the percentage change in projected readership between year 1 and year 5.

Price elasticity of demand

All businesses are interested in how sensitive demand is to a change in price. It is important to know what will happen to demand for a product if the price is either increased or reduced. This topic includes how to calculate the price elasticity of demand (PED) for a product and how to calculate the change in demand following a change in price, given the PED. Consideration will then be given to the impact of a change in price on demand and revenue, and how to calculate income elasticity of demand.

The calculation for PED concerns the relationship between price and demand, and is part of the marketing section of A Level specifications. The formula required for this calculation is:

$$\text{Price elasticity of demand} = \frac{\text{percentage change in quantity demanded}}{\text{percentage change in price}}$$

The shorthand version of the formula is: $\dfrac{\%\Delta Qd}{\%\Delta P}$

The answer for price elasticity of demand is negative because the variables move in opposite directions – as the price falls, demand increases and vice versa. However, when discussing price elasticity of demand, you can ignore the negative sign and focus on the absolute number.

If the number is less than 1 (ignoring the minus sign) then the PED is **inelastic**, i.e. the percentage change in quantity demanded is less than the percentage change in price. This means an increase in price will increase total revenue, but a reduction in price will lead to a fall in total revenue.

If the number is greater than 1 (ignoring the minus sign) then the PED is **elastic**, i.e. the percentage change in quantity demanded is greater than the percentage change in price. This means an increase in price will lead to a fall in total revenue, whereas a reduction in price will lead to an increase in total revenue.

WORKED EXAMPLES

Piri-Piri

As a national chain of Portuguese-style fast food restaurants, Piri-Piri is always concerned that its prices ensure that revenue is maximised. As part of a recent promotion, the price of its set menu was reduced from £20 to £17. Sales increased in the North West region from 9,000 per month to 10,080. The marketing director is keen to know the price elasticity of demand and therefore whether total revenue for the region has increased.

Calculate the PED for the North West region and estimate the change in total revenue.

$$\text{Price elasticity of demand} = \frac{\text{percentage change in quantity demanded}}{\text{percentage change in price}}$$

Percentage change in quantity = (change ÷ original) × 100
$$= (1{,}080 \div 9{,}000) \times 100 = 12$$

Percentage change in price = (change ÷ original) × 100
$$= (-3 \div 20) \times 100 = -15$$

PED = 12 ÷ −15 = (−)0.8

Estimate: the PED is inelastic (less than 1) and therefore total revenue will fall when price is reduced.

Proof: 9,000 set meals at £20 = £180,000; 10,080 meals at £17 = £171,360

Christmas cards

If the PED for charity Christmas cards is estimated to be $(-)2.0$, a fall in the price by 15% will lead to which of the following?

a A 30% decrease in the quantity demanded

b A 7.5% increase in the quantity demanded

c A 30% increase in the quantity demanded

d A 7.5% decrease in the quantity demanded

Formula: $\dfrac{\%\Delta Qd}{\%\Delta P}$ $\qquad \dfrac{?}{-15} = -2 \qquad ? = -2 \times -15 = 30$

The answer is **c**, a 30% increase in the quantity demanded.

PRACTICE QUESTIONS

1 A manufacturer of traditional boiled sweets sells a wide range of products in packs of five large plastic jars. The price for each pack varies according to the cost of sugar, which has to be purchased on a monthly basis from a UK sugar refining company.

The figures for the first four months of the year are given here:

Month	January	February	March	April
Price	20.00	20.30	20.60	21.63
Quantity sold	950	935	900	891

Calculate the PED for the packs of sweets as the price rose between March and April.

2 If ticket prices for the next FA Cup final are increased by 6% and the PED is -0.5, what is the anticipated change in the quantity demanded and what would be the expected change in total revenue?

> **REMEMBER:** Although a negative PED will be expressed with the minus sign (e.g. –0.2 or –2), you do not have to worry about it when doing the calculation.

STRETCH YOURSELF

Income elasticity of demand (YED)

Income elasticity of demand measures the relationship between a change in quantity demanded and a change in income. The basic formula for calculating the coefficient of income elasticity is:

$$\text{Income elasticity of demand} = \frac{\text{percentage change in quantity demanded of good A}}{\text{percentage change in consumers' real income}}$$

Most goods will have a positive income elasticity of demand, i.e. as income rises, so does demand. Luxury goods are likely to be more income elastic (number greater than 1) and necessities are likely to be income inelastic (number less than 1).

If average disposable incomes increase by 2.5% and sales of hot tubs increase from 6,000 per month to 6,240, what is the YED?

Time series analysis

This is a method used to predict future trends – usually sales figures – based on past data. An analysis of past figures can show whether there has been a positive upward trend or a downward trend. This is useful information for business managers involved in strategic decision making, such as deciding whether to expand production capacity.

Moving averages

Using past figures, such as sales data, an average can be calculated which reveals an underlying trend and smooths out fluctuations. The decision about how many time periods to include in a moving average will depend on the time periods over which the data have been gathered. For example, quarterly data would require a four-period moving average. Annual data are often analysed by using a three-period moving average. Calculating a three-period moving average for annual data involves adding the first three annual figures together and dividing by three. The next calculation involves dropping the first year's value and adding on the fourth, and so on.

WORKED EXAMPLE: CALCULATING MOVING AVERAGES

Below are annual the sales figures for Beddy Byes Bed Company.

Year	1	2	3	4	5	6	7	8
Sales (£000)	600	550	620	730	700	750	800	850

Calculate a three-year moving average and comment on the trend.

First the sales for years 1–3 are added together and divided by three:

$$\frac{600 + 550 + 620}{3} = \frac{1,770}{3} = 590$$

To create a moving average, the first year is now dropped and year 4 is added:

$$\frac{550 + 620 + 730}{3} = \frac{1,900}{3} = 633$$

The process continues until the final results are:

Year	1	2	3	4	5	6	7	8
Trend		590	633	683	727	750	800	

The moving average results – also known as the trend – show a steady increase in sales.

PRACTICE QUESTION

1 In June last year Cheesecakes Ltd launched a major promotional campaign in the hope of gaining more customers and the managing director, Katie Butcher, wants to know whether it was successful. Here are the company's most recent sales figures:

Jan	Feb	March	April	May	June	July	Aug	Sept	Oct	Nov	Dec	Jan
5,400	7,020	7,560	6,480	6,480	6,480	7,250	7,520	7,610	7,320	7,580	8,100	5,600

Calculate a three-month moving average. Does there seem to be an upward or downward trend in sales?

Notice that the moving average in the worked example above is shown at the centre of the three years. For example, the moving average for years 1–3 is shown under the year 2 figure. This is because it is an odd number. Calculating moving averages based on an even number of values is a little more complex.
For example, if using quarterly data, placing the average result somewhere between quarters 2 and 3 might be misleading. In cases where even-number moving averages are required, a technique called **centring** is used.

WORKED EXAMPLE: CENTRING

The sales data in the following table relate to Pop Squash, a business selling soft drinks to retailers in a holiday resort. Complete columns 4, 5 and 6 by calculating a four-period moving total and moving average. Assess the trend.

1 First, a four-period moving total is calculated in column 4 by adding the sales values for years 1–4, then years 2–5, years 3–6 etc.:

120 + 140 + 190 + 130 = 580;
140 + 190 + 130 + 130 = 590

2 To complete column 5, two quarters' moving totals are added together to produce the eight-quarter moving total:

580 + 590 = 1,170;
590 + 610 = 1,200

1	2	3	4	5	6
Year	Quarter	Sales revenue (£000)	4-quarter moving total (£000)	8-quarter moving total (£000)	Centred moving average (trend) (£000)
2011	1	120			
	2	140			
	3	190			146.25
	4	130	580		150.00
2012	1	130	590	1,170	156.25
	2	160	610	1,200	163.75
	3	220	640	1,250	167.50
	4	160	670	1,310	168.75
2013	1	130	670	1,340	172.50
	2	170	680	1,350	176.25
	3	240	700	1,380	181.25
	4	170	710	1,410	187.50
2014	1	160	740	1,450	191.25
	2	190	760	1,500	193.75
	3	250	770	1,530	
	4	180	780	1,550	

3 The figures in column 5 are then divided by 8 to give the centred average (column 6). This 'centres' the first average *(1,170 ÷ 8 = 146.25)* alongside quarter 3 of 2011. This is because although eight values have been added, these are taken from just five quarters of data and the mid-point of five values is the third. The next average would be *(1,200 ÷ 8)* 150, which would be placed next to quarter 4 of 2011 (the mid-point of the next five values).

The trend values, because they have been averaged, eliminate most of the seasonal variation and they show a steady increase in sales over the years in question.

There may also be cyclical variations in data that occur over time periods longer than the four seasons in a year. The most likely cause of these is the business cycle, which causes the sales of most products to fall during a recession but rise during a boom.

PRACTICE QUESTION

2 The sales of Ahmed's convenience store were recorded over a four-week period. The shop is open only five days per week. The owner of the store has started to undertake a short-term forecasting exercise.

		Sales	5-day moving total	Moving average
Week 1	Tues	32		
	Wed	38		
	Thurs	40		43
	Fri	50		44
	Sat	55	215	
Week 2	Tues	37	220	
	Wed	40	222	
	Thurs	42	224	
	Fri	53	227	
	Sat	55		

		Sales	5-day moving total	Moving average
Week 3	Tues	40		
	Wed	44		
	Thurs	50		
	Fri	60		
	Sat	70		
Week 4	Tues	44		
	Wed	48		
	Thurs	52		
	Fri	65		
	Sat	72		

Complete the five-day moving total and moving average.

Assess whether the sales trend for Ahmed's shop is rising or falling.

Variations from trend

If the figures for the moving average are plotted on a graph, it is possible to see that fluctuations have been smoothed out and an underlying trend revealed. If a line of best fit is now drawn (visually) then this can be projected into the future to show the forecasted trend values. This 'extrapolated' trend line gives forecasts which can then be adjusted, for example by the average seasonal variation for each quarter.

The variation of the actual data from the trend data is important to calculate in order to be able to make more accurate future forecasts (see the worked example on seasonal variation below).

The actual sales data for the time period are compared to the trend data using the formula:

Variation = actual data − trend data

This formula can be seen in operation in the seasonal variation worked example below. For example, in 2011 quarter 3: actual (190) − trend (146.25) = 43.75

Seasonal variations

The process of variation analysis is particularly useful for businesses that are interested in the impact of seasonal trends. By predicting future results using trend analysis and then adjusting these with seasonal variations, more accurate sales forecasts can be obtained.

 WORKED EXAMPLE

Looking back to the worked example for Pop Squash, plot the centred moving average results on a graph and compare them with the actual sales figures.

If the trend is identified on the graph by line of best fit, this line can then be extended (extrapolated) for future trend sales levels to be forecast.

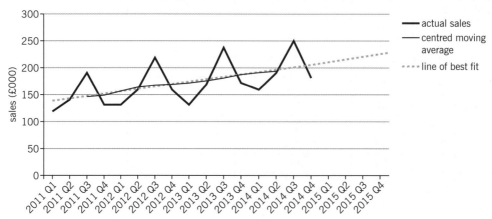

However, a more accurate forecast can be made by adjusting the future trend by the expected seasonal variation of each quarter. The following table shows how the seasonal variation (SV) for each quarter is calculated as: SV = column 3 − column 6

The seasonal variation in column 7 is calculated using: variation = actual data − trend data

For example: 2011 Q4 variation = 130 − 150 = −20

Then the average seasonal variation in column 8 is obtained for each quarter using the formula:

$$\text{Average seasonal variation} = \frac{\text{total of all seasonal variations for one quarter}}{\text{number of results for this quarter}}$$

For example for quarter 3, this is:

$$\frac{\text{2011 Q3 SV} + \text{2012 Q3 SV} + \text{2013 Q3 SV}}{3} = \frac{43.75 + 52.5 + 58.75}{3} = 51.67$$

1	2	3	4	5	6	7	8
Year	Quarter	Sales revenue (£000)	4-quarter moving total (£000)	8-quarter moving total (£000)	Centred moving average (trend) (£000)	Seasonal variation (£000)	Average seasonal variation (£000)
2011	1	120					
	2	140					
	3	190			146.25	43.75	51.67
	4	130	580		150.00	−20.00	−15.40
2012	1	130	590	1,170	156.25	−26.25	−33.30
	2	160	610	1,200	163.75	−3.75	−4.60
	3	220	640	1,250	167.50	52.50	51.67
	4	160	670	1,310	168.75	−8.75	−15.40
2013	1	130	670	1,340	172.50	−42.50	−33.30
	2	170	680	1,350	176.25	−6.25	−4.60
	3	240	700	1,380	181.25	58.75	51.67
	4	170	710	1,410	187.50	−17.50	−15.40
2014	1	160	740	1,450	191.25	−31.25	−33.30
	2	190	760	1,500	193.75	−3.75	−4.60
	3	250	770	1,530			
	4	180	780	1,550			

This result can be used as follows. Using the line of best fit above, the trend forecast for quarter 3 in 2015 is £220,000. This trend forecast is then adjusted to give an actual sales forecast by adding on the average seasonal variation for quarter 3:

£220,000 + £51,670 = £271,670

This is the sales forecast for quarter 3 of 2015.

PRACTICE QUESTION

3 Use the sales data given below to calculate the centred moving average trend, the quarterly seasonal variation and the average seasonal variation for each quarter.

Plot a graph of the moving average (trend) data and add a line of best fit.

Forecast a trend sales value for quarter 2 in year 4.

Seasonally adjust this trend forecast by the average seasonal variation for quarter 2.

Year	1				2				3			
Quarter	1	2	3	4	1	2	3	4	1	2	3	4
Sales	450	260	250	400	475	280	240	410	480	295	250	425

Analysing data: averages

This topic will consider techniques used to analyse data, such as averages and measures of dispersion, including standard deviation.

A lot of the data collected by businesses is not in a form that is useful for decision making. It is necessary for it to be reorganised so that it can be used more effectively. To find out the most common outcome from data, the central tendency or average is calculated.

For example, a firm might want to know its:
- average sales per month or per year
- average inventory turnover
- average number of days lost through sickness.

There are three measures of 'average' or 'central tendency': arithmetic mean, median and mode.

The arithmetic mean

This is what most people think of as an average. It is calculated by adding the value of all items and dividing by the number of items.

The formula is:

Mean = sum of items ÷ number of items

 WORKED EXAMPLE

Company A

Below are the sales figures for the company for a 40-week period (000s).

8	10	12	12	14	10	8	9	10	12
8	12	14	12	12	9	9	8	12	12
8	10	12	12	14	10	8	8	10	12
10	12	8	8	9	8	10	12	14	10

Taking the first four weeks, the mean would be:

$$\frac{8 + 10 + 12 + 12}{4} = \frac{42}{4} = 10.5$$

The calculation of the mean by this method is time consuming. To save time, and improve accuracy, the frequency with which each number occurs can be calculated. This is called the **frequency distribution** and is calculated by multiplying the quantity sold by the frequency, then adding up these totals and dividing by the total frequency.

WORKED EXAMPLE

Company A

The frequency distribution for the sales figures on page 14 is shown below.

Quantity sold (000s)	Frequency	Quantity × frequency
8	10	80
9	4	36
10	9	90
12	13	156
14	4	56
TOTALS	**40**	**418**

Dividing the quantity × frequency by the frequency gives the mean.

$418 \div 40 = 10.45$, i.e. 10,450

The company may use this average sales figure when forecasting revenue and profit.

However, because this calculation uses all the data, it can be distorted by extreme values, which may be misleading. For example, if during four weeks the sales had been 24,000 rather than 14,000, the total would have been $458 \div 40 = 11.45$, i.e. 11,450. If this figure is used to forecast revenue and profit, the predictions may be over–optimistic.

PRACTICE QUESTION

1 Below are the sales figures for Tops for Sport for a 20-week period at the start of last year.

9.25	9.5	9.75	9.75	10	10.25	11	11	10.75	10.75
10.75	10.5	10.25	10	9.75	9.75	9.75	9.5	9.5	9.25

a Calculate the arithmetic mean correct to two decimal places for the first four weeks and the last four weeks of the sales period.

b Calculate the mean using the frequency distribution method for the 20-week sales period.

The median

This is the middle number in a set of data. When numbers are put in numerical order, the median is the number in the middle. For example, the median of 1, 2, 3, 4, 5 would be 3, whereas the median of 1, 2, 3, 4 would be 2.5 because this is the halfway point.

The formula for calculating the position of the median in a set of ordered data is as follows:

$(n + 1) \div 2$ (for an odd number of values) or $n \div 2$ (for an even number of values), where $n =$ the number of values or total frequency.

WORKED EXAMPLE:
FROM GROUPED FREQUENCY DATA

Company A

1 Quantity sold (000s)	2 Frequency	3 Cumulative frequency	4 Quantity (1) × frequency (2)
8	10	10	80
9	4	14	36
10	9	23	90
12	13	36	156
14	4	40	56
TOTALS	**40**		**418**

There are 40 values in this table of sales and therefore the median is $(n \div 2) = 40 \div 2 = 20$. The 20th value in the table has a sales value of £10,000. This is because, according to the cumulative frequency data, the 15th to 23rd values occurred at this level of sales.

This method of calculating the average is not distorted by extreme values, but it does have limited statistical value.

PRACTICE QUESTION

2 Use the frequency table created in practice question 1 to calculate the median.

The mode

This is the value that appears most frequently. The mode is not affected by extreme values and is easy to calculate. However, it does not take into account all the values and might be misleading if it is used as a measure of the average. This situation is made more likely when there is more than one mode within a set of data.

WORKED EXAMPLE

Company A
The frequency distribution of the sales figures is shown below.

Quantity sold (000s)	Frequency	Quantity × frequency
8	10	80
9	4	36
10	9	90
12	13	156
14	4	56
TOTALS	**40**	**418**

The mode is **12,000 units** sold because this is the most frequent, occurring 13 times.

PRACTICE QUESTION

3 Use the frequency table created for practice question 1 to calculate the mode.

Dispersion

Having calculated the average, a business will probably be interested in knowing how typical this figure is. If there is a great distance between the highest and lowest values in the data spread, this is a **wide dispersion** and it is likely that the average may not be particularly useful. If, however, the data have a **narrow dispersion**, then the average will have a stronger relationship to the rest of the data and may be of more value to the business.

There are different ways to calculate the spread of data and the following table will be used to explain these methods.

Production figures for Potter's Brogues (pairs)												
Month	Jan	Feb	Mar	Apr	May	Jun	Jul	Aug	Sep	Oct	Nov	Dec
	140	136	160	130	130	125	120	100	160	140	136	155

If these output figures are put in ascending value order they are:

100	120	125	130	130	136	136	140	140	155	160	160

Range

This is the easiest method. It is the difference between the highest and lowest values in the range. In the example above, this would be $160 - 100 = 60$.

The problem with this method is that the range can be distorted by extreme values. In this example, the output of 100 pairs of shoes in August was exceptional due to extreme weather conditions. Without this value the range would have been $160 - 120 = 40$. For this business, this figure is much more realistic.

Interquartile range

This method looks at the range within the central 50% of the data provided: that is, it ignores the bottom and top 25% (quarters). This reduces the impact of extreme values.

To calculate the interquartile range, the data must be arranged in ascending value order, as shown above. The value which is one-quarter of the way along is the first quartile and the value which is three-quarters of the way along is the third quartile. The difference between these two figures gives us the interquartile range.

WORKED EXAMPLE: INTERQUARTILE RANGE

This is a simple example to illustrate the principle using the following formula:

1st quartile $(Q1) = (1 \times n) \div 4$

where $n =$ the total number of values.

In the example above there are 12 values (months) so:

$n \div 4 = (1 \times 12) \div 4 = 3$

This calculation shows the value below which 25% of all the figures fall.

To calculate the third quartile, the following formula is used:

3rd quartile $(Q3) = (3 \times n) \div 4 = (3 \times 12) \div 4 = 36 \div 4 = 9$

This calculation shows the value above which 25% of all the figures fall.

Using the ascending value order, the third item is 125 and the ninth is 140 so the interquartile range is:

$140 - 125 = 15$

This gives a much narrower value than the range calculated above.

> **REMEMBER:** If there are very large amounts of data, other values may be used instead of quartiles, such as 50% values. In the Potter's Brogues example, this would be 50% of 12 (total number of values) = 6. The sixth value is **136** pairs of shoes.

Mean deviation

The range and the interquartile range only consider the spread between two figures in a set of data. However, in each set of data there are many figures and each will deviate from **the mean**. This could happen for many reasons:

- sales vary on a month-by-month basis – for example, depending on weather conditions
- market research conducted in two or more geographical regions gives different results
- the output of a new machine may vary as initial problems are ironed out.

 WORKED EXAMPLE

Potter's Brogues

In the Potter's Brogues example, the arithmetic mean for output of shoes is calculated as:

Sum of items ÷ number of items = 1,632 ÷ 12 = 136

The deviation from this mean is shown in the table here.

The mean deviation produces one figure, by averaging the differences in all values from the mean. It is normal to ignore the plus and minus signs and use the formula:

Mean deviation $= \sum(x - \bar{x}) \div n$

where:

$(x - \bar{x})$ is the difference between each value (in this case, output) and the mean (ignoring the minus sign)

n is the number of values

$\sum$ is the mathematical notation for the 'sum of'.

The mean deviation for the monthly output figures for Potter's Brogues is 150 ÷ 12 = 12.5.

This is the average deviation of all values from the mean. The larger the mean deviation, the wider the spread or dispersion.

The main problem with this method is the removal of the plus and minus signs. This is dealt with by two further methods, the **variance** and the **standard deviation**.

Months	Output (x)	Deviation $\bar{x}$ ($\bar{x}$ − 136)
Jan	140	4
Feb	136	0
Mar	160	24
Apr	130	−6
May	130	−6
Jun	125	−11
Jul	120	−16
Aug	100	−36
Sep	160	24
Oct	140	4
Nov	136	0
Dec	155	19
		Total = 150

 PRACTICE QUESTION

4 Below is a table giving the monthly production figures for Blackcountry Blades for last year. Output is given in 000s of units.

Month	Jan	Feb	Mar	Apr	May	Jun	Jul	Aug	Sep	Oct	Nov	Dec
Output	40	46	52	54	54	52	58	56	54	56	42	36

Using this information, calculate:

a the range

b the interquartile range

c the mean deviation.

Variance and standard deviation

As we have seen, there are several methods by which a firm can calculate dispersion. However, they all lack accuracy and therefore value to a business. The **variance** allows the firm to consider the average of a spread of all data from the mean.

WORKED EXAMPLE

Potter's Brogues

Using the information about Potter's Brogues, the plus and minus figures can be removed by squaring the numbers rather than ignoring the signs. This is shown in the fourth column.

Months	Output (x)	Deviation (x − 136)	Deviations squared
Jan	140	4	16
Feb	136	0	0
Mar	160	24	576
Apr	130	−6	36
May	130	−6	36
Jun	125	−11	121
Jul	120	−16	256
Aug	100	−36	1,296
Sep	160	24	576
Oct	140	4	16
Nov	136	0	0
Dec	155	19	361
			Total = 3,290

The variance is calculated by: $\left(\sum(x - \bar{x})^2\right) \div n = 3{,}290 \div 12 = 274.17$

The original figures were expressed in units (000s) of production, but the variance figures are expressed in units squared. To return to the original units it is necessary to find the square root of the variance. This is known as the **standard deviation**:

$\sqrt{274.17} = 16.56$

PRACTICE QUESTION

5 Using the table from the previous practice question, calculate the deviations squared, the variance and the standard deviation for this business.

Using standard deviation: normal distribution

The normal distribution is a statistical model that tells a business what the expected range of outcomes from a particular sample might be. It is used in market research when large-scale sampling is carried out, and in quality control when the firm wants to know what range of results to expect. A **normal distribution curve** shows all the possible outcomes and the frequency at which they will occur. It is bell shaped and symmetrical about the **mean value**.

The shape of a normal distribution curve is determined by the spread of data and this spread can be measured by the use of standard deviations.

Whatever the spread of the normal distribution curves, they have particular features:
- the curve is symmetrical about the mean
- the mean, mode and median of the distribution are equal
- 50% of all values lie either side of the mean value
- the curve can be divided into three standard deviations (SDs) either side of the mean.

Thus nearly all results will lie with + or − 3 SDs of the mean. A small proportion will lie outside this range, but this is so small that businesses are usually not concerned about it in practice.

The normal distribution curve shows the results for +/−

1 SD reflect approximately 68% of the results; those for +/−

2 SD reflect approximately 95% of the results; and those for +/−

3 SD reflect approximately 99% of the results.

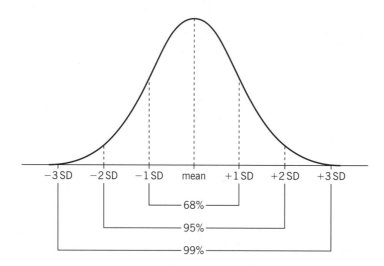

![checkmark] **WORKED EXAMPLE**

Cheesecakes

The company has decided to launch a new product and, after six months of heavy promotion, has asked a market research company to discover whether or not customers know about the new cheesecake. Before the promotion the company would expect 50% of those asked to recognise a product, but after the promotion it would expect the figure to be higher. If the market research company asks 500 customers, what results might the company expect in order to get a measure of whether or not the promotion has been successful?

Firstly, the range needs to be established by calculating the mean and the standard deviation for this particular distribution.

The mean for normal distribution can be calculated by:

$$\text{Mean} = n \times p$$

where n = the sample size and p = the probability of the event occurring.

For the market research that Cheesecakes commissioned for the new product, $n = 500$ and $p = 0.5$. Therefore the mean = 250.

The standard deviation can be calculated using the formula:

$$1\,\text{SD} = \sqrt{npq}$$

where n = the sample size, p = the probability of the event occurring and q = the probability of the event not occurring.

$n = 500; p = 0.5; q = 0.5$

Therefore $1\,\text{SD} = \sqrt{500 \times 0.5 \times 0.5} = \sqrt{125} = 11$ (approx.)

The full range of results can be $+$ or $-$ $3\,\text{SD}$ from the mean, where $2\,\text{SD} = 22$ and $3\,\text{SD} = 33$. Therefore the range for this normal distribution curve will be:

$250 +$ or $- 33 = 217$ to 283

Using the normal distribution curve shown on page 20, the firm can predict that:

68% of all results will show that between 239 and 261 people will recognise the new cheesecake, given a mean of 250 ($250 - 1\,\text{SD} = 239; 250 + 1\,\text{SD} = 261$).

95% of all results will show that between 228 and 272 people will recognise the new cheesecake ($250 - 2\,\text{SD} = 228; 250 + 2\,\text{SD} = 272$).

99% of all results will show that between 217 and 283 people will recognise the new cheesecake ($250 - 3\,\text{SD} = 217; 250 + 3\,\text{SD} = 283$).

These percentages are usually referred to as confidence levels.

PRACTICE QUESTION

6 Computer Support Ltd offers telephone support for businesses that have bought computer systems. The company's USP is that it offers better customer service than the manufacturer. An important feature of its system is that it expects to answer telephone calls within 18 seconds. If the time taken to answer the telephone is normally distributed around this 18-second mean with a standard deviation of three seconds, what is the probability that a random customer's call will be answered in less than 12 seconds?

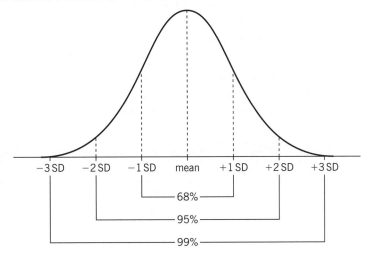

Total revenue

Total revenue is defined as the average price per item multiplied by the quantity sold. This can be shown as a formula:

Total revenue = average selling price × quantity sold

A business will need to know how much revenue it can expect to receive as this is an important part of the profit calculation.

WORKED EXAMPLES

Jones' Jewels

Marissa Jones' jewellery business sells a range of earrings, necklaces and bracelets at an average price of £50. In an average month she sells 300 items of jewellery. What is the total revenue earned by Marissa's business?

Total revenue = average selling price × quantity sold
= £50 × 300 = £15,000

Claire's Cheesecakes

This business sells a variety of cheesecakes from Claire Watson's kitchen. These are sold by the slice to local cafés and restaurants at an average price of £4.50 per slice. There are 12 slices in each cheesecake. In June, the business sold a total of 120 cheesecakes. Calculate the total revenue for June.

Revenue per cheesecake = £4.50 × 12 = £54

Total revenue for June = £54 × 120 = £6,480

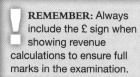

REMEMBER: Always include the £ sign when showing revenue calculations to ensure full marks in the examination.

PRACTICE QUESTIONS

1 Shiny Windows is a window cleaning business run by Gill Smith. She charges customers £8 to clean up to six windows and £12 if it is more than six. In a normal month Gill cleans 40 residences for £8 and 50 residences for £12. What is her total revenue per month?

2 Craig Potter runs a business making exclusive handmade brogue shoes in Nottingham. Each pair is made to measure and they retail at an average price of £140. Below is a table of sales for the first six months of last year. Complete the table by calculating the total revenue for each month.

	January	February	March	April	May	June
Sales	20	15	30	35	35	20
Price (£)						
Total revenue (£)						

Calculating price or quantity from revenue

Sometimes you may be given information about revenue and will be expected to calculate the average price or the quantity sold.

✓ WORKED EXAMPLES

Classy Cleaners

This franchise business claims that new franchisees can earn on average £1,600 per month in the first year. The rate for cleaning is £20 per hour with an average cleaning time of two hours per client. Assuming there are four weeks per month, how many clients would a franchisee need to achieve **per week** to achieve this income per month?

Total revenue ÷ price = quantity sold (or number of customers)

Averge price per client = £20 × 2 = £40

£1,600 ÷ £40 = 40 clients per month, 10 clients per week

Guy's Cakes

Guy Clarke's business makes unique celebration cakes targeting the growing male market. He has been operating for a very busy six months and has calculated that his total revenue is £12,000. Each cake has an individual price based on the requirements of his clients, but Guy and his accountant want to know what the average price per cake is. Guy has made 24 cakes in the first six months of running his business. What is the average price per cake?

Total revenue ÷ quantity sold = average price

£12,000 ÷ 24 = £500 per cake

> **!** **REMEMBER:** Always include the formula when doing any calculation.

? PRACTICE QUESTIONS

3 Daisy Priest runs a successful business, Daisy's Dog Nursery, in the Long Eaton area of Nottingham. She collects dogs from busy owners in the morning and looks after them all day, returning them home at 6pm at the latest. Daisy's friend Sharon lives in Derbyshire and wants to open a similar business in her local area, where lots of business people with dogs live. Sharon knows that Daisy earned £1,200 last month (four weeks) and charges £20 per day to her clients, who use the service for five days per week. If she charges the same price per day, how many clients would Sharon need to achieve the same revenue as Daisy?

4 The Breaston Garden Seat Company manufactures high-quality garden seats. The total revenue for last year was £3,250,000 with annual sales of 25,000 seats. Calculate the selling price of each garden seat.

Total and unit costs

Total, fixed and variable costs

Another element of the calculation of profit is total costs. This is calculated by adding total fixed costs and total variable costs. Fixed costs are defined as those which do not change as the business changes its level of output or sales. Variable costs are defined as those which change directly with output or sales.

✓ WORKED EXAMPLES

Jones' Jewels

Marissa Jones' jewellery business has fixed costs of £400 per month. Marissa has calculated that variable costs are £2 per item for materials and labour. She sells an average of 300 items of jewellery per month. Calculate her total costs.

Total costs = total fixed costs + total variable costs
= £400 + (£2 × 300) = £1,000 per month

Claire's Cheesecakes

This business sells a variety of cheesecakes from Claire Watson's kitchen. The fixed costs are £300 per month and the variable cost is £2.50 per slice of cheesecake for materials, labour, packaging and transport. There are 12 slices in one cheesecake and in an average month 120 of the cheesecakes are sold. Calculate the total costs.

Total costs = total fixed costs + total variable costs
Total variable costs = (£2.50 × 12) × 120 = £3,600
Total costs = £300 + £3,600 = £3,900

? PRACTICE QUESTIONS

1   Shiny Windows is a window cleaning business run by Gill Smith. She has very low fixed costs of £50 per month for insurance. Her variable costs are also very low at £3 per customer, which is mostly petrol. Calculate Gill's total costs if she completes 90 jobs per month.

2   Craig Potter runs a business making exclusive handmade brogue shoes in Nottingham. Each pair is made to measure and the variable costs are £70 per pair including labour and materials. Fixed costs are £450 per month. Use the table below to calculate his total costs per month.

	January	February	March	April	May	June
Sales	20	15	30	35	35	20
Fixed costs						
Variable costs						
Total costs						

Calculating average/unit costs from total costs

Businesses often want to know what the average cost of producing one item is. This enables them to make a direct comparison with price to indicate whether or not profit can be made. Obviously this will depend on the number of items sold. Average costs are also known as unit costs and are calculated by dividing total costs by the number of units sold:

Average (unit) costs = total costs ÷ number of items produced

WORKED EXAMPLE

Classy Cleaners

This franchise business has fixed costs of £700 per month including a royalty to the franchisor of £6,000 per year. The total variable costs are £640 per month. Calculate the average cost per visit if the franchisee visits 10 customers, four times per month.

Average (unit) cost = (total fixed costs + total variable costs) ÷ number of visits to customers
= (£700 + £640) ÷ 40 = £33.50

PRACTICE QUESTION

3 Daisy Priest runs a successful business, Daisy's Dog Nursery, in the Long Eaton area of Nottingham. She collects dogs from busy owners in the morning and looks after them all day, returning them home at 6pm at the latest. Daisy's fixed costs are £37.50 per week and her variable costs are £5 per day per dog. Assuming there are four weeks per month, calculate Daisy's total costs per month and her average (unit) cost per dog per day if she has three dogs every day for five days per week.

Calculating changes in variable costs

The number of customers is a vital component of total cost calculations. If the number of customers changes, this will have an impact on costs as well as revenue.

WORKED EXAMPLE

Guy's Cakes

Guy Clarke's business makes unique celebration cakes targeting the growing male market. The fixed costs are £1,000 per month and the variable costs per cake are £150 including labour. Guy makes four cakes per month. Calculate his total costs and unit costs per cake.

Total costs = total fixed costs + total variable costs
= £1,000 + (£150 × 4) = £1,600

Unit costs = total costs ÷ sales

£1,600 ÷ 4 = £400 per cake

A competitor opens in the same area as Guy and undertakes an aggressive advertising campaign. Guy finds that sales fall to three cakes per month. How does this impact on Guy's costs?

Total costs = total fixed costs + total variable costs
= £1,000 + (£150 × 3) = £1,450

Unit costs = £1,450 ÷ 3 = £483 per cake

Guy's unit costs rise by £83 per cake.

PRACTICE QUESTION

4 Richard Thompson runs a small business making bespoke trailers for a range of customers in his local area. His fixed costs are £12,690 per year and his variable costs are £375 per trailer. Richard builds an average of 60 trailers per year.

a Calculate his total costs per year and unit costs per trailer.

b Richard is offered cheaper components by a supplier whom he meets at a trade fair in Birmingham. This will reduce his variable costs by 20%.

Calculate the impact that buying components from the new supplier will have on Richard's total and unit costs.

Profit, loss and profit margin

Once the revenue has been calculated and the total costs are known, a business is able to calculate the level of profit or loss achieved.

Businesses are also interested in the percentage of the final selling price which is profit. This is known as the profit margin.

Finally, a business wants to know the relationship between the quantity of products sold and the profits made. Under normal conditions, if a business sells more products, then the profits made will also increase, as long as it does not have to reduce the price in order to increase sales (see pages 8–9 for price elasticity of demand).

Profit and loss

Profit or loss is defined as total revenue minus total costs.

WORKED EXAMPLES: PROFIT CALCULATIONS

Jones' Jewels
Marissa Jones' jewellery business sells a range of earrings, necklaces and bracelets at an average price of £50. In an average month she sells 30 items of jewellery. She has calculated her sales revenue per month as £1,500. She has also calculated that her total costs per month are £1,000. What is Marissa's profit?

Profit = total revenue − total costs = £1,500 − £1,000 = £500 per month profit

Claire's Cheesecakes
This business sells a variety of cheesecakes from Claire Watson's kitchen. These are sold by the slice to local cafés and restaurants at an average price of £4.50 per slice. There are 12 slices in each cheesecake. In June, the business sold a total of 120 cheesecakes. The fixed costs are £300 per month and the variable cost is £2.50 per slice for materials, labour, packaging and transport. Calculate the profits for each month from January to June by completing the following table (answers are shown in bold).

June: profit = total revenue − total costs (fixed costs + variable costs) = £6,480 − £3,900 = £2,580

April and May can be calculated in a similar way.

January: total revenue = £4.50 × 1,200 = £5,400; total variable costs = £2.50 × 1,200 = £3,000; profit = £5,400 − (£300 + £3,000) = £2,100

February and March can be calculated in a similar way.

	January	February	March	April	May	June
Price (£)	4.50	4.50	4.50	4.50	4.50	4.50
Sales (slices)	1,200	1,560	1,680	1,440	1,440	1,440
Total revenue (£)	**5,400**	**7,020**	**7,560**	6,480	6,480	6,480
Fixed costs (£)	300	300	300	300	300	300
Variable costs @ £2.50 (£)	**3,000**	3,900	4,200	3,600	3,600	3,600
Total costs (£)	**3,300**	**4,200**	**4,500**	3,900	3,900	3,900
Profit (£)	**2,100**	**2,820**	**3,060**	2,580	2,580	2,580

> **REMEMBER:** It is very important to remember that the formula is total revenue minus total costs, not the other way round. This is a very common mistake.

WORKED EXAMPLE: A LOSS-MAKING BUSINESS

Classy Cleaners
This franchise business claims that new franchisees can earn on average £1,600 per month in the first year. The rate for cleaning is £20 per hour with an average cleaning time of two hours per client. The business has fixed costs of £700 per month including a royalty to the franchisor of £6,000 per year. The variable costs are £16 per visit. Kathy bought a Classy Cleaners franchise and in the first month she attracted five clients per week. Assuming there are four weeks per month, calculate Kathy's profit or loss for this month.

Profit = total revenue − total costs

Total revenue = price × sales = £40 × (5 × 4) = £800

Total costs = fixed costs + variable costs = £700 + (£16 × 20) = £1,020

Kathy has made a loss of: £800 − £1,020 = £220

PRACTICE QUESTIONS

1 Shiny Windows is a window cleaning business run by Gill Smith. She has calculated that her revenue for July was £920, and that her total costs were £320. Calculate Gill's profit for July.

2 Craig Potter runs a business making exclusive handmade brogue shoes in Nottingham. Each pair is made to measure and he sells them at an average price of £140. The variable costs are £70 per pair including labour and materials. Fixed costs are £450 per month. Use the table below to calculate the profit made by the business in the first six months of the year.

	January	February	March	April	May	June
Price (£)	140					
Sales	20	15	30	35	35	20
Total revenue (£)			4,200			
Fixed costs (£)	450					
		1,050				
Total costs (£)					2,900	
Profit (£)						

Profit margin

This is calculated in the following ways:
- profit per unit as a percentage of selling price
- the difference between the selling price and the unit cost.

WORKED EXAMPLE: CALCULATING THE PROFIT MARGIN

Daisy's Dog Nursery

Daisy Priest runs a successful business looking after the pets of working people. She charges £20 per day to her three clients and her unit costs are £7.50 per dog per day. Calculate Daisy's profit margin.

Profit margin = selling price − unit cost = £20 − £7.50 = £12.50

Expressed as a percentage of the selling price: (£12.50 ÷ 20) × 100 = 62.5% profit margin.

PRACTICE QUESTION

3 Guy Clarke's business is making unique celebration cakes targeting the growing male market. He has calculated that the average price for a cake is £500 and that his unit costs are £400 if he makes four cakes per month. What is his profit margin expressed as a value in pounds and as a percentage of the selling price?

STRETCH YOURSELF

What would happen to the profit margin of Guy's Cakes in the following circumstances?

a His unit costs rise by 10% as a result of rising world food prices.

b His average price per cake then rises by 40%.

Budgets

Completing budgets

Budgets are forward financial plans for revenue/income, costs/expenditure and profits. You may be expected to complete a budget for a business or make changes to an existing plan.

Profit is calculated using the following formula:

Profit = sales revenue (or total sales) − total costs

You may have to add up more than one source of revenue to calculate total revenue. Alternatively you may have to calculate different costs by deducting from total costs, as shown in the following worked example.

WORKED EXAMPLE

Beautiful Books

This independent bookshop in Shrewsbury is owned by Sarah Elbourne. She has started to produce her budget for the next four months. Complete the budget targets in the table below by filling in the spaces **a–d**.

	March	April	May	June
Sales revenue	6,000	b	6,750	7,000
Labour costs	4,000	4,000	4,000	3,500
Rent paid	1,000	1,000	1,000	1,000
Other costs	a	800	900	1,100
Total costs	5,600	5,800	5,900	d
Profit	400	700	c	1,400

a 600 (5,600 − 5,000)　　**b** 6,500 (5,800 + 700)　　**c** 850 (6,750 − 5,900)　　**d** 5,600 (7,000 − 1,400)

PRACTICE QUESTION

1　The Soul Food Restaurant started business six months ago and budgeted costs for the sixth month were £5,000 wages; £1,000 rent and £550 food. Actual costs were £5,750 wages; £1,000 rent and £500 food. The owners have produced the following budgets which need to be completed.

	Month 1	Month 2	Month 3	Month 4	Month 5	Month 6
Revenue from food	2,750	3,000	3,000	3,250	3,500	3,500
Revenue from drinks	3,000	3,500	3,500	3,750	3,750	4,000
Total revenue	a	6,500	6,500	7,000	7,250	7,500
Wages	3,000	3,500	3,500	4,000	5,000	5,000
Rent	1,000	1,000	1,000	1,000	1,000	1,000
Food and drink	450	475	475	500	525	550
Total costs	4,450	4,975	c	5,500	6,525	6,550
Profit	b	1,525	1,525	d	725	e

Amending budgets

Sometimes businesses need to change their budget targets when they obtain more accurate information or when costs change unexpectedly. You will probably have to calculate percentage changes using the following formulae:

Percentage increase = original × (1 + the percentage★)

★e.g.　5% increase = 1.05; 15% increase = 1.15

Percentage decrease = original × (1 − the percentage★★)

★★e.g.　5% decrease = 0.95

Alternatively, the change might be a whole number and require a simple addition or subtraction.

WORKED EXAMPLE

Hotter UK

Hotter is a UK footwear retailer. Its budgeted sales for month 1 were 2,000 units at £60, giving total revenue of £120,000. Its actual sales were 1,800 units. The owners were determined to make their budgets as accurate as possible and commissioned some market research. As a result they have decided to amend their sales budget by reducing it by 10% while raising their prices by an average of 8%. Complete the new retail budget in the table below.

	Sales	Average price	Total revenue
Month 1			

Sales

To calculate a 10% reduction, multiply by 0.9: $2,000 \times 0.9 =$ **1,800**

Alternatively, divide the original sales figure by 100 and multiply by 10. This number is then subtracted from the original sales figure: $(2,000 \div 100) \times 10 = 200$; then $2,000 - 200 =$ **1,800**

Average price

To calculate an increase of 8%, multiply by 1.08: $60 \times 1.08 =$ **£64.80**

Or $(60 \div 100) \times 8 = 4.8$; then £60 + £4.80 = **£64.80**

Total revenue

$£64.80 \times 1,800 = £116,640$

PRACTICE QUESTION

2   Simon Jones of Jones Trading estimated the start-up budgets for the first six months of his new business, based on his own limited experience and the advice of his best friend and business partner, Geoff. However, when they showed their targets to a local business adviser, she suggested significant changes to make the budgets more realistic.

Complete the following table showing the adjustments to the original budget, then calculate the change to the total costs and profit budget.

	Original budget (£)	Recommended change	New budget (£)
Sales revenue	26,000	20% decrease	
Wages	6,000	No change	
Rent	1,500	5% increase	
Materials	5,000	30% increase	
Other costs	2,500	£375 increase	
Total costs			
Profit			

Variances

Calculating favourable and adverse variances

Once a budget has been completed or amended, the next stage is to compare the actual figures with those in the budget. The formula for this process is:

Variance = actual figures − budgeted figures

For revenue or profit variances, if the variance is negative this is an **adverse** variance because it means that the actual revenue or profit is **lower** than the budgeted figure, which is a negative outcome for the business.

For cost variances, if the variance is negative this is a **favourable** variance because it means that the actual cost is **lower** than the budgeted figure, which is a positive outcome for the business.

✓ WORKED EXAMPLES

Hotter UK

Hotter is a UK footwear retailer. Its budgeted sales for one month were 2,000 units at £60. Its actual sales were 1,800 units at £62.50.

The accounts department calculated a sales variance of which of the following?

a £120,000 favourable

b £120,000 adverse

c £7,500 adverse

d £7,500 favourable

Variance = actual sales − budgeted figure

= £112,500 − £120,000 = −£7,500

As the actual sales revenue is less than the budgeted figure, the variance is adverse.

£7,500 adverse

Books are Beautiful

This is the budget for a small bookshop. Complete the variance analysis.

	Budget	Actual	Variance	Adverse/ favourable
Sales revenue	6,000	5,820		
Labour costs	4,000	3,790		
Rent paid	1,000	1,100		
Other costs	600	640		
Profit	400	290		

Sales revenue = 5,820 − 6,000 = 180 adverse

Labour costs = 3,790 − 4,000 = 210 favourable

Rent paid = 1,100 − 1,000 = 100 adverse

Other costs = 640 − 600 = 40 adverse

Profit = 290 − 400 = 110 adverse

> **REMEMBER: Adverse variance** is when the actual figure is a negative outcome for the business.
>
> **Favourable variance** is when the actual figure is a positive outcome for the business.

PRACTICE QUESTIONS

1 The Soul Food Restaurant business started six months ago and budgeted costs for the sixth month were £5,000 wages; £1,000 rent and £550 food. Actual costs were £5,750 wages; £1,000 rent and £500 food.

The accounts department calculated the total cost variance as which of the following?

a £700 adverse

b £700 favourable

c £750 favourable

d £750 adverse

2 This is the budget for a business start-up that has been trading for six months. Complete the variance analysis.

	Budget	Actual	Variance	Favourable/ adverse
Sales revenue	28,500	32,000		
Wages	14,000	16,000		
Stocks	7,000	8,000		
Other costs	6,000	5,500		
Profit	1,500	2,500		

 REMEMBER: To calculate the variance the formula is:

Variance = actual figure − budgeted figure

Cash flow

Calculating, constructing and amending cash-flow forecasts

A cash-flow forecast **predicts** cash inflows (revenue) and cash outflows (costs of raw materials, labour, etc.) per month over a given time period.

A cash-flow statement is a financial document which shows the cash inflows and cash outflows for a business over a given time period in the past.

The monthly cash outflows are deducted from the cash inflows, leaving a net cash flow for each month. This is added to the opening bank balance for that month to give a forecasted closing bank balance for that month. This figure becomes the opening bank balance for the next month.

✓ WORKED EXAMPLE

Beautiful Books

The bookshop has been operating very successfully and benefits from an annual literary festival in Shrewsbury every August. The owner, Sarah Elbourne, has produced a cash-flow forecast for the coming year.

What are the missing figures at points **i** and **ii** in the extract from the forecast shown here? Select your answer from the options **a**–**d** below.

Cash inflows	July	August
Sales	8,500	12,500
Cash outflows		
Materials	1,200	1,600
Labour costs	3,000	3,500
Rent and other costs	1,200	1,200
Total payments	5,400	ii
Net cash flow	3,100	6,200
Opening bank balance	1,500	4,600
Closing bank balance	i	10,800

Point **i**

a 10,000 negative   **c** 4,600 positive
b 1,600 negative   **d** 7,700 positive

Answer: **c**
Net cash flow + opening balance = 3,100 + 1,500 = 4,600

Point **ii**

a 5,000   **b** 6,300   **c** 6,000   **d** 18,800

Answer: **b**
Materials + labour costs + rent and other costs = 1,600 + 3,500 + 1,200 = 6,300

? PRACTICE QUESTIONS

1 The table below shows a partially completed cash-flow forecast for the Soul Food Restaurant. Calculate the closing balance for November. Show your working.

Item	October (£)	November (£)
Opening balance	1,000	
Inflows		
Sales revenue	2,750	3,000
Outflows		
Wages	3,000	3,500
Food and drink	450	475
Rent and other costs	1,000	1,000
Net cash flow		
Closing balance		

> **REMEMBER:** Be prepared for a range of layouts and terms, and have the confidence to cope with this. For example, cash inflows may be called 'receipts' and cash outflows may be called 'payments'.

2 Jack Gordon started his business, Imports to Glow, importing glow sticks from China when he was still at school. He produces monthly cash-flow forecasts for his business. What will be the new forecasted **net cash flow** for July if Jack now expects cash sales to be 25% higher and materials to cost 20% more?

Item	July (original) (£000)	July (new) (£000)
Cash inflow		
Cash sales	2.5	
Payments from debtors	0.5	
Total cash inflow	3	
Cash outflow		
Labour	0.2	
Materials	0.75	
Overheads	0.25	
Total cash outflows	1.2	
NET CASH FLOW	1.8	

> **!** REMEMBER: **Net cash flow** is always total inflows/receipts *minus* total outflows/payments for a particular month.

STRETCH YOURSELF

Construct a cash-flow forecast

From the information given, construct a forecast for two months.

Month 1
Opening balance £0.2m
Revenue £1.2m
Wages £0.3m
Materials £0.7m
Overheads £0.25m

Month 2

Revenue £1.25m
Wages £0.3m
Materials £0.9m
Overheads £0.3m

Item	Month 1	Month 2
Opening balance		
Inflows		
Sales revenue		
Outflows		
Wages		
Materials		
Overheads		
Total outflows		
NET CASH FLOW		
Closing balance		

Costing methods

'Costing' means measuring the costs of a business activity. This provides managers with financial information to help with decision making. Calculating product costs can also aid the marketing department when setting prices.

Contribution

Contribution is one of the most important concepts in Business Studies.

Contribution is the difference between selling price and variable costs. The difference between these two figures is the amount that will be **contributed** towards paying the total fixed costs of the business and achieving a profit. Therefore it should be expected that the contribution will be a positive number.

Unit contribution

This is the contribution to fixed costs and profit of one unit sold. The formula for calculating the unit contribution is:

Unit contribution = selling price − variable cost per unit

The formula may also be presented as:

Unit contribution = selling price − direct costs per unit

✔ WORKED EXAMPLE

Jones Jewels

Marissa Jones makes jewellery and estimates that her customers pay on average £50 per item. She also estimates her variable costs, including the metal, jewels and packaging, at £20 per item. According to her accountant, last month Marissa sold 25 items and achieved total revenue of £1,125. Her total costs were £1,000, of which £600 were fixed costs, including renting premises, equipment and gallery space.

Calculate the contribution made per unit based on Marissa's estimates.

Calculate the contribution based on her accountant's figures.

How accurate are Marissa's estimates?

Marissa's estimates: Selling price − unit variable costs = 50 − 20
$$= £30 \text{ unit contribution}$$

Accountant's figures: Price = total revenue ÷ sales = 1,125 ÷ 25 = £45

Unit variable costs = total variable costs ÷ sales = 400 ÷ 25
$$= £16$$

Unit contribution = price − unit variable costs = 45 − 16
$$= £29 \text{ contribution}$$

Marissa's estimates were accurate to within £1.

? PRACTICE QUESTION

1   Sophia Olagunju of Creative Curtains makes curtains using fabric selected by her clients from high-quality London department stores. She works from home, which means that Sophia does not have very high fixed costs, although she did have to borrow money to buy her industrial sewing machine and other specialist equipment. Sophia charged her last customer £500 for two sets of fully lined,

full-length curtains. The variable costs, including the fabric, heading tape and thread, were £325.

Sophia has just received an order from a local hotel for 10 sets of curtains for bedroom refurbishments. Sophia hopes that this will lead to further orders if the hotel likes the quality of her work. She has quoted a price of £2,500 for the curtains and calculates that her total variable costs will be £1,600.

Calculate the unit contribution from Sophia's last order.

Calculate the unit contribution from the hotel order.

Make an initial analysis of whether supplying the hotel would be a good strategy for Sophia.

Total contribution

When a firm has a large order, it can calculate the total contribution it makes towards the fixed costs and profit.

Total contribution = sales revenue − total variable cost

> **REMEMBER:** You may see variations on this formula depending on the examination board used. These include:
>
> Total contribution = total revenue − total direct costs
>
> Total contribution = (selling price − unit direct costs) × output or units sold

✓ WORKED EXAMPLE

Guy's Cakes

Guy Clarke has expanded his product portfolio to include a range of 'Power Snacks'. The variable costs are 45p per bar. A national chain of gyms wants to place an order for 50,000 snack bars at a price of £1.20 each. Calculate the total contribution of this order.

Total contribution = total revenue − total variable costs

Total revenue = 50,000 × £1.20 = £60,000

Total variable costs = 50,000 × £0.45 = £22,500

Total contribution = £60,000 − £22,500 = £37,500

or

Total contribution = (selling price − unit variable costs) × output or units sold

Total contribution = (£1.20 − £0.45) × 50,000 = £0.75 × 50,000 = £37,500

PRACTICE QUESTION

2 Tracey Walshaw runs a successful business printing on sports tops for a wide range of clubs around the East Midlands. Her fixed costs last year were £52,000 and the sports tops cost Tracey £4.25 each. Other variable costs such as power, printing and labour added a further £3.59 per top. Tracey charges £10.20 per sports top, and had sales of 9,500 units. Calculate Tracey's total contribution for last year.

> **REMEMBER:** You need to make sure that the figures you have are matched; you cannot mix unit figures with total figures.
>
> Example: Calculate total contribution from the following information.
> Total revenue (TR) = £9,000; variable cost per unit = £6.00; sales = 1,000 units.
>
> The first step is to calculate total variable costs (TVC):
>
> $£6.00 \times 1,000 = £6,000$
>
> Total contribution = $TR - TVC$ = £9,000 − £6,000 = £3,000
>
> Similarly, to calculate unit contribution the first step is to calculate price:
>
> Total revenue ÷ sales = £9,000 ÷ 1,000 = £9
>
> Unit contribution = price − VC per unit = £9 − £6 = £3

Change in contribution

You may also be asked to calculate the change in contribution when two time periods are compared. The formulae given below can be used for this type of question.

Change in contribution = contribution in time period 2 − contribution in time period 1

or

Change in contribution = additional revenue − additional variable costs

WORKED EXAMPLE

Handmade Trailers

Richard Thompson retired from his job two years ago and started a business making bespoke trailers. After the first year Richard was disappointed by the number of sales and decided to reduce his average price by 12% to £748. As a result, sales rose from 48 to 60 trailers. Use the table to calculate the change in total contribution between year 1 and year 2.

Financial period	Selling price (£)	Sales	Variable costs* per item (£)
1	850	48	375
2	748	60	325

*Variable costs include raw materials and labour.

Total contribution = (selling price − unit variable costs) × sales

Year 1 total contribution: (£850 − £375) × 48 = £475 × 48 = £22,800

Year 2 total contribution: (£748 − £325) × 60 = £423 × 60 = £25,380

The change in total contribution is: £25,380 − £22,800 = £2,580

PRACTICE QUESTION

3 Craig Potter, the owner of Potter's Brogues, calculates that a pair of brogues generates an average contribution of £70 per sale. Last year in the period April to June, 90 pairs were sold, but due to changes in fashion, Craig believes that sales will be down by 20%. Assuming no change in price or costs, calculate the change in total contribution that Craig might expect for the period April to June this year.

STRETCH YOURSELF

The Breaston Garden Seat Company

This single-product company manufactures high-quality garden seats and has the following financial structure:

Variable costs per unit: materials £30; labour £32; overheads £10

Fixed costs: £420,000 per year

Price: £130 per seat

Annual sales: 25,000

Capacity: 31,250

A national garden centre chain has enquired about whether it can buy 6,000 seats to sell as 'own brand' items. It is willing to pay £85 per seat. By accepting the order, the Breaston Garden Seat Company will incur an additional £10,000 set-up costs.

Based on the information provided, should the order be accepted?

Absorption or full costing

This can be used to find out the cost of manufacturing a specific product or the cost of operating an individual cost/profit centre of a business. Profit centres are departments or divisions of a business to which costs and revenue can be allocated. Absorption/full costing allows the company to calculate the profit made by each centre, which then enables comparisons to be made about the performance of different parts of the business.

Absorption/full costing allocates both the direct costs and the indirect costs (or overheads) to profit centres. This is **not** the case with contribution costing, which does not allocate indirect costs (see pages 34–7).

In absorption/full costing, indirect costs of the business have to be 'absorbed' into or allocated to each cost/profit centre. There are several ways of making this allocation decision and the following example illustrates the process.

WORKED EXAMPLE:
ALLOCATION BASED ON PERCENTAGE OF TOTAL DIRECT COSTS

The Patchwork Company manufactures patchwork quilts. It produces two sizes. Total indirect costs are £600,000 per year. The direct costs of production are:

Product:	Single quilt	Double quilt
Direct costs:	£400,000	£600,000

Allocate the indirect costs of the company as a percentage of total direct costs.

Calculate the total costs for each duvet size.

Total direct costs are £400,000 + £600,000 = £1,000,000.

The single quilt accounts for 40% of total direct costs: (£400,000 ÷ £1m) × 100

The double quilt accounts for 60% of total direct costs: (£600,000 ÷ £1m) × 100

Indirect costs can be allocated on this 40 : 60 basis:

Single quilt: £600,000 × 0.4 = £240,000

Double quilt: £600,000 × 0.6 = £360,000

This means that total costs for each product are:

Single quilt: £400,000 + £240,000 = £640,000

Double quilt: £600,000 + £360,000 = £960,000

Total costs can then be deducted from the revenue for each quilt to calculate the profit made by each.

The problem with this 'simple' method of allocating indirect costs is that it might be inaccurate. If, for example, both quilt sizes take up the same factory floor space, the double quilt has been allocated a proportion of indirect costs in excess of this proportion. This could result in misleading management data, for example on profit levels, and inappropriate management decisions, for example on pricing levels.

PRACTICE QUESTION

4   The Queen's Head Hotel is a country hotel that has divided its business into three profit centres: Accommodation, Restaurant and Bar. Details of the direct costs are given below. The hotel incurs indirect costs of **£600,000** per year.

	Accommodation	Restaurant	Bar
Direct labour costs:	£50,000	£160,000	£30,000
Other direct costs:	£30,000	£140,000	£90,000

a   Calculate the total direct costs for each profit centre.

b   Calculate the total direct costs for the business.

c   Allocate the indirect costs of the hotel as a percentage of total direct costs.

d   Calculate the full costs for each profit centre.

e   Is there an alternative way of allocating indirect costs that could be more accurate?

Apportion method

This is a way of trying to allocate each indirect cost more accurately. Here are some options:

- rent: allocated by floor space used as proportion of total floor space
- human resource department costs: allocated by the proportion of total employees in each department
- IT administration and maintenance costs: allocated by the value of computers used per department.

WORKED EXAMPLE: APPORTIONING OVERHEADS

Napiers of York

This independent department store has three departments and each one operates as a profit centre.

This helps the business to monitor which departments are most successful. The senior accountant has suggested that new methods of apportioning indirect costs should be considered, rather than the percentage of total direct costs currently used.

Below are the direct costs of operating each profit centre and some other information relating to the costs of the business. The value of overheads last year was **£800,000**: £320,000 in rent, £280,000 in warehousing costs and £200,000 for administration and other indirect costs.

	Womenswear	Childrenswear	Electricals	Total
Direct costs	300,000	100,000	600,000	1,000,000
Workers employed	6	3	6	15
Floor space used	600 m²	300 m²	300 m²	1,200 m²

Apportion indirect costs as follows: rent by floor space; administration costs by number of employees; and warehousing costs by direct cost proportion.

	Womenswear	Childrenswear	Electricals
Rent	50% of total floor space 50% of rent = £160,000	25% of floor space 25% of rent = £80,000	25% of floor space 25% of rent = £80,000
Administration costs	40% of total employees 40% of admin costs = £80,000	20% of total employees 20% of admin costs = £40,000	40% of total employees 40% of admin costs = £80,000
Warehouse costs	30% of direct costs 30% of warehouse costs = £84,000	10% of direct costs 10% of warehouse costs = £28,000	60% of direct costs 60% of warehouse costs = £168,000

If the revenue earned last year by each profit centres was:
Womenswear £400,000
Childrenswear £300,000
Electricals £600,000

then the profit made would be:

Womenswear: £400,000 − £160,000 − £80,000 − £84,000 = £76,000

Childrenswear: £300,000 − £80,000 − £40,000 − £28,000 = £152,000

Electricals: £600,000 − £80,000 − £80,000 − £168,000 = £272,000

These recorded profit figures would be different if the indirect costs were apportioned using other methods.

STRETCH YOURSELF

Blackcountry Blades

This company manufactures wiper blades, in packs of 100 blades, for the automotive industry and operates three profit centres:
P1: Cars and vans, P2: Lorries and buses and P3: Other vehicles.

Factory time and direct costs per pack of blades are shown in the table below. There are four types of indirect cost: rent (£48,000), marketing (£72,000), utilities (£96,000) and administration (£16,000). Rent is apportioned according to factory time used and administration is apportioned according to the labour input of each profit centre. The other two indirect costs are apportioned equally between all three profit centres. Blackcountry Blades produces **4,000 blade packs** in each profit centre per year.

Profit centre	Labour (£)	Materials (£)	Other (£)	Factory time (minutes)
P1	4	6	2	30
P2	4	8	4	60
P3	8	4	4	30
Total	16	18	10	120

Calculate the full cost **per pack of blades** for each department using the absorption method outlined on pages 37–8.

Special order decisions

When asked to consider whether or not to continue with the production of a product or whether to accept a specific order, it is important to consider only **contribution costing** (see pages 34–7) and not absorption (or full) costing. As a general rule, as long as a special order makes a positive contribution it is likely to be accepted, and if an existing product makes a contribution it is likely to be continued with.

Sometimes a business receives an order that is 'out of the ordinary'. For example, it may be from a new customer or it may have very specific requirements which may involve changes to normal production processes. The business must decide whether or not to accept the order.

If absorption costing is used and existing indirect costs are apportioned to the special order then it might appear to make a loss. Should the order be rejected? 'No' is almost always the correct response because the **indirect/overhead costs** have to be paid anyway. It is much more accurate to ignore existing indirect costs and just calculate the contribution (revenue less direct costs) that the special order will make.

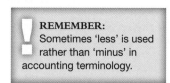

REMEMBER:
Sometimes 'less' is used rather than 'minus' in accounting terminology.

WORKED EXAMPLE

Handmade Trailers

This growing business makes bespoke trailers. The fixed costs are £12,690 per year, the average direct costs per trailer are £325 and the average price per trailer is £748. Richard, the owner of the business, sells 60 trailers per year. This gives him an operating profit as follows:

$$\text{Total revenue} = £748 \times 60 = £44,880$$

$$\text{Total cost} = £12,690 + (£325 \times 60) = £32,190$$

$$\text{Operating profit} = £44,880 - £32,190 = £12,690$$

Richard receives an enquiry for an order for 10 trailers, but the client is only willing to pay £400 per trailer. Should Richard accept the order? Richard has the production capacity to make these trailers.

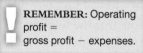

REMEMBER: Operating profit = gross profit − expenses.

$$\text{Contribution} = (\text{selling price} - \text{unit direct costs}) \times \text{output} = (£400 - £325) \times 10$$

This gives a total contribution of £750, which adds to the profit of the business.

Note that only the direct costs are taken into consideration, because **the indirect costs have already been covered by the original 60 trailers sold each year**.

If all 70 trailers were sold at the lower price of £400, Handmade Trailers would make a loss:

$$\text{Total revenue} = £400 \times 70 = £28,000$$

$$\text{Total cost} = £12,690 + (£325 \times 70) = £35,440$$

$$\text{Profit/loss} = £28,000 - £35,440 = -£7,440$$

So, the additional order should only be accepted if the other 60 trailers can be sold for the average price of £748.

$$60 \text{ trailers at } £748 = £12,690 \text{ profit}$$

$$10 \text{ trailers at } £400 = £750 \text{ profit}$$

$$\text{Total profit} = £13,440$$

PRACTICE QUESTION

5 Harry George runs a small business making shopping bags from recycled materials. He sells the bags to a range of retail stores for a price of £4. At the moment he is producing 100,000 bags per year, although there is capacity to produce a further 30,000. Harry has been approached by a new customer who wants 20,000 bags, but is only willing to pay £2.50 each.

Harry's cost structure is shown in the table below. Calculate the annual contribution figure and the profit made without the new order. On financial grounds, should Harry accept the order? Show your workings.

Materials	Per bag	£0.70
Direct labour costs	Per bag	£1.60
Other direct costs	Per bag	£0.50
Indirect/overhead costs		£80,000

Break-even output and the margin of safety

Calculating the break-even output

This is the level of output or sales at which the business is making just enough to cover all costs, but not making any profit.

The calculation is made using the following formula:

$$\text{Break-even output} = \frac{\text{fixed costs of the business}}{\text{contribution per unit}\star}$$

★See page 34 for how to calculate contribution.

Note that the solution to the calculation is not expressed in financial terms; it is a level of output or sales to be achieved.

✓ WORKED EXAMPLE

Tops for Sport

Tracey Walshaw runs a business which prints team details on sports tops for a wide range of clubs. Last month she sold 600 tops at a price of £12.00 per item. Her fixed costs last month were £2,000 and the average variable costs were £7.00 per top. Calculate the break-even level of sales for last month.

$$\text{Break-even output} = \frac{\text{fixed costs}}{\text{contribution per unit}} = \frac{£2,000}{£12 - £7} = \frac{£2,000}{£5} = 400 \text{ tops}$$

? PRACTICE QUESTIONS

1   Craig Potter makes handmade shoes in Nottingham with an average selling price of £140 per pair. His fixed costs for the coming year are £5,600 and his average variable costs last year were £70. If there is no change to his costs, how many pairs of shoes will Craig have to make and sell in order to break even this year?

2   Marissa Jones makes jewellery from her home in the Lake District. She has created the following summary of her financial transactions for last year. What was the break-even level of sales for her business? Explain your answer by showing your workings.

Item	£
Fixed costs	6,000
Total revenue	18,000
Contribution per unit	40
Variable costs	6,000

3   Richard Thompson's business, Handmade Trailers, has fixed costs of £12,690 per year. He used to sell trailers for £850 with an average variable cost of £375 per trailer. He now sells trailers at an average price of £748 and the average variable cost is £325 per trailer. Calculate the break-even level of output per year before and after his change in price.

4   Gill Smith is a self-employed window cleaner. She has fixed costs of £280 per month and average variable costs of £2 per client. If she charges £10 on average, how many clients does she need per month to break even?

> **! REMEMBER:** When expressing break-even, your calculation should be given in terms of units of output or sales, not a financial value.

STRETCH YOURSELF

The Soul Food Restaurant

The head chef has calculated that he needs to serve 50 meals per day at an average price of £20 to break even. Fixed costs are approximately £200 per day. What are the average variable costs per meal at the break-even output?

Margin of safety

The margin of safety compares the firm's current level of output and break-even. Obviously a business will have a margin of safety only if it is producing or selling **above** the level required to break even.

You may be given details of sales and be asked to calculate not only the break-even output but also the margin of safety achieved over a given time period. To do this you will require the following formula:

> Margin of safety = current sales (output) − break-even sales (output)

WORKED EXAMPLE

Tops for Sport

Tracey Walshaw calculates that her break-even level of sales is 400 tops per month. Last month she sold 600 tops at a price of £12.00 per item. Calculate her margin of safety.

> Margin of safety = current sales − break-even sales
> = 600 − 400 = 200

PRACTICE QUESTIONS

5 Gill Smith is a self-employed window cleaner. She has fixed costs of £280 per month and average variable costs of £2 per client. If she charges £10 on average, and has 40 clients on her books, what is her margin of safety?

6 Craig Potter has decided to move to new premises in Nottingham, closer to more exclusive shops such as Paul Smith. By increasing his average selling price of £155 per pair, Craig hopes to cover his higher fixed costs of £8,000 and his average variable costs of £75. Calculate his break-even level of sales and his margin of safety if he achieves sales of 112 pairs per year.

7 Marissa Jones of Jones' Jewels has experienced rising variable costs giving her a variable cost per unit of £30. She still hopes to achieve a margin of safety of 150 items if she increases her prices to an average of £60 and sells 280 items next year. Is Marissa correct in her hopes if her fixed costs remain at £6,000? Show your working to justify your answer.

8 Richard Thompson's business has fixed costs of £12,690 per year. He used to sell 48 trailers per year for £850 with an average variable cost of £375 per trailer. He now sells 60 trailers per year at an average price of £748 and the average variable cost is £325 per trailer. Has his margin of safety increased or decreased?

Break-even charts

Constructing, interpreting and adjusting break-even charts

Break-even charts give a visual indication of the level of output or number of sales a business will need to cover all its costs. They can also be used to show the margin of safety and the level of profit or loss achieved for a given level of sales or output. Due to time constraints, it is very unlikely that you will be asked to construct a complete chart under examination conditions. However, there is no such limitation for applied business courses, where the ability to construct a break-even chart may be beneficial.

WORKED EXAMPLE: CONSTRUCTING AND USING A BREAK-EVEN CHART

Potter's Brogues

Craig Potter is planning to open another branch of his business in London. He intends to run the new business himself, leaving his assistant to manage the one in Nottingham. After completing both primary and secondary research, Craig has made the following forecasts:

Fixed costs: **£1,600** per month

Average variable costs: **£80** per pair of shoes

Price: **£180** per pair of shoes

Craig knows from experience that the maximum number of pairs of shoes he can make per month is 40.

Draw a break-even chart for Craig's business.

The first stage is to mark the scales on the two axes. Craig knows that the maximum level of output is 40, so this means that the scale on the horizontal axis runs from 0 to 40. The vertical scale records the costs and revenues. We can assume that revenue will be the highest figure, so you can calculate Craig's total revenue if he sells 40 pairs of shoes per month (£180 × 40) as **£7,200**. Thus the scale on the vertical axis runs from 0 to 7,200.

Add the fixed cost line to the chart. This is a horizontal line at **£1,600** because these are the costs that do not change with output. This is shown on the chart on page 45.

Next draw the total cost line. This is the addition of fixed costs and variable costs at each level of output. You can calculate total costs at zero output and at maximum output and then join these two points together with a straight line. At zero output total costs are **£1,600**, i.e. just the fixed costs are payable if nothing is being produced. At maximum production, the total costs will be **£1,600** plus **£80** (average variable cost per pair of shoes) × **40** (maximum output) = **£1,600** + **£3,200** = **£4,800**. This is shown on the chart on page 45.

Finally add the total revenue line to the graph, using the same approach of zero output (sales) and the maximum output of 40 pairs of shoes. You calculated the maximum revenue to be **£7,200**, so the total revenue line starts at zero and ends at **£7,200**.

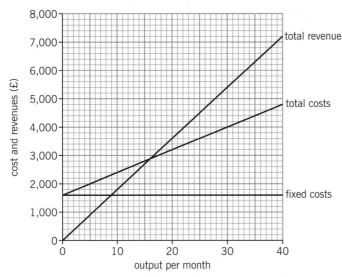

Once the chart is drawn and labelled, you can answer questions based on it.

1 Identify the number of sales that Craig Potter must achieve to break even.

16 pairs of shoes (i.e. where the total cost line intersects the total revenue line).

Note that some margin for error is always given with such questions.

2 If Potter's Brogues sells 30 pairs of shoes in the first month, how much profit or loss will be made?

At output 30, total revenue is £5,400 and total cost is £4,000, therefore profit is £1,400.

3 If Potter's Brogues sells 10 pairs of shoes in the first month, how much profit or loss will be made?

At output 10, total revenue is £1,800 and total cost is £2,400, therefore it makes a loss of £600.

4 If Potter's Brogues sells 25 pairs of shoes per month, what is the margin of safety?

Margin of safety = current output (sales) − break–even output/sales

Current output = 25 and break-even output = 16, therefore the margin of safety is 9 pairs of shoes.

Note that the own-figure rule will apply if earlier calculations are required for an answer. In other words, any mistakes made earlier will not be penalised again.

PRACTICE QUESTION

1 Helen Rose runs a part-time business arranging short-term lets for people with property in Wales. She charges 15% commission on property, which achieves an average rent of £600 per month. Helen has fixed costs of £300 per month and variable costs of £30 per property per month, and believes that the maximum number of properties she can manage is 10 per month.
Draw and label a break-even chart for Helen's business. State the break-even number of properties she needs to let and identify it on the chart with the letter A. If Helen lets eight properties per month, what is her profit? If Helen lets seven properties per month, what is her margin of safety?

Changes in the break-even level of output

Most firms want to reduce the break-even level of output or sales because this suggests that the business can begin to make a profit sooner. In order to achieve this, one of the elements of the calculation must improve:

- Fixed costs are reduced.
- Variable/unit/direct costs are reduced.
- Price is increased.

WORKED EXAMPLE: CHANGES IN COSTS AND/OR REVENUE

The Soul Food Restaurant

The Soul Food Restaurant has been running very successfully for one year and needs to move to bigger premises. This will increase the fixed costs but will allow more meals to be served per day. At the moment the fixed costs are £6,000 per month and the variable cost per meal is £6. A meal at the Soul Food Restaurant costs £12 on average. The business is open 20 days per month for lunch and evening meals.

At the new premises, the variable costs and the price per meal will stay the same to begin with; however, the fixed costs will rise to £9,000 per month. The owners hope that the better and bigger location will bring in more customers.

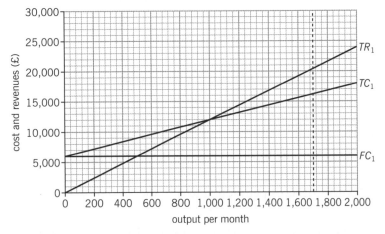

1 Amend the break-even chart above to show the cost changes.

2 State the change in the break-even point.

3 Show the change in profit based on 1,700 meals per month. Label the original profit as AB on the chart, and the new profit as AC.

4 If the average number of meals per month is 1,700, explain how the margin of safety has changed.

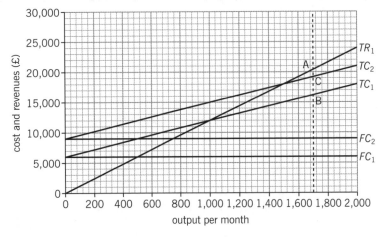

1 The new fixed cost line is at £9,000 and shown as FC_2.

If we divide TR_1 by the average price per meal, we can assume that the maximum output is 2,000 meals (24,000 ÷ 12).

To add the new total cost line, calculate the total cost at output 2,000 meals per month, then draw a straight line from the start point of £9,000.

Total cost at 2,000 meals = fixed cost + total variable cost;
£9,000 + (2,000 × £6) = £9,000 + £12,000 = £21,000. The new total cost line is shown as TC_2 and it starts at the same place as the new fixed cost line.

2 The original break-even point was 1,000 meals (i.e. where the line TR_1 intersects TC_1).

The new break-even point is 1,500 meals (i.e. where the line TR_1 intersects TC_2).

The change in break-even is that an additional 500 meals must be sold per month to break even.

3 The profit at 1,700 meals has changed from AB to AC.

4 The margin of safety has changed from 1,700 − 1,000 = 700 to 1,700 − 1,500 = 200

PRACTICE QUESTION

2 Sarah Elbourne of Beautiful Books has started to organise a series of literary events for groups such as schools, adult education colleges and women's organisations in the Shrewsbury area as part of an initiative to encourage reading amongst children and adults. So far she has organised 20 events for which she charges £400, from which she expects to make a profit. Sarah aims to have at least one famous author attending each function and this means that her variable costs are £200 to cover their expenses. Her fixed costs are £2,000 per year. However, as the price of petrol continues to rise, her guest authors have asked for an increase in expenses and so her variable costs have risen to £300 per event.

Show the change to the total cost line on the graph below by drawing a new total cost line.

Identify the new break-even point.

Forecast the new profit/loss if Sarah runs 16 events per year. Show this on the chart using the labels A and B.

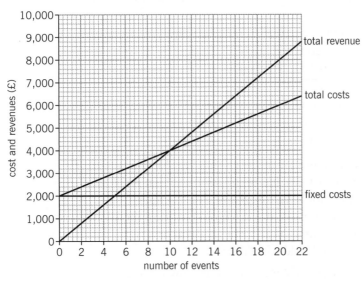

> **REMEMBER:** When asked to read off an existing break-even chart, use a ruler to ensure accuracy.

Depreciation: straight line and reducing balance

Depreciation is concerned with the value of fixed assets as shown in the company accounts. Fixed assets, in contrast to current assets, are those intended for continued use over a period of time. It is assumed that most fixed assets will lose value over their lifetime (property and land being the main exceptions), and so this must be shown when valuing them for accounting purposes. Two methods of calculating depreciation will be discussed here: the straight line method and the reducing balance method.

Straight line depreciation

This is the easiest way of calculating depreciation and requires three pieces of information:
- the purchase price of the asset (also known as the historic cost)
- the expected life of the asset
- the residual value (if any) of the asset at the end of its expected life. This is the estimated value of the asset at the end of its useful life.

The following formula can then be used to calculate the annual figure which is to be included in the financial accounts of the business to express the loss in value of the asset as it is used:

$$\text{Annual depreciation charge} = \frac{\text{purchase price} - \text{residual value}}{\text{estimated life of the asset}}$$

✓ WORKED EXAMPLE

Company A

The operations director has proposed that the firm buys a new machine for **£50,000** that will benefit the business by increasing productivity and enabling an increase in potential output for the next **five years**. The machine is estimated to have a residual value of £5,000.

Calculate the first year's depreciation for Company A using the straight line method.

$$\text{Annual depreciation charge} = \frac{\text{purchase price} - \text{residual value}}{\text{estimated life of the asset}}$$
$$= \frac{£50,000 - £5,000}{5}$$
$$= \frac{£45,000}{5}$$
$$= £9,000$$

This figure represents the depreciation figure for the asset each year for the five years of its life. Each year, the depreciated value of the asset is shown on the balance sheet. This is called the **net book value** (NBV). For example, for Company A, at the end of year 2 the value of the machine shown on the balance sheet would be £32,000. At the end of year 5 the asset will have a value of £5,000: the same as the estimated residual value.

The net book value (NBV) of the asset over its five-year life is shown in the table and figure on the next page.

Extract from the balance sheets of Company A

	Annual depreciation	NBV
Asset at cost £50,000		
31 December Yr 1	£9,000	£41,000
31 December Yr 2	£9,000	£32,000
31 December Yr 3	£9,000	£23,000
31 December Yr 4	£9,000	£14,000
31 December Yr 5	£9,000	£5,000

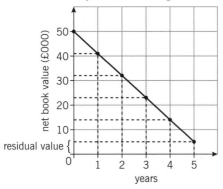

Graph showing the NBV of the machine over its five-year lifetime (straight line method)

> **REMEMBER:** No cash is involved in the depreciation process. It is simply a 'book' transaction to try and ensure that assets are valued accurately. This means that the company's financial statements are not misleading.

PRACTICE QUESTION

1 The board of directors of Company B has to consider an investment opportunity that includes the acquisition of a new vehicle costing £60,000. The vehicle is estimated to have a useful life of five years and a residual value of £8,000. Use the straight line method of depreciation to calculate the annual rate of depreciation.

STRETCH YOURSELF

Company B

The board of directors and the company auditors want to know the net book value at the end of year 3 and year 5 of the vehicle's life. Use your answer from the practice question above and the information provided to perform the necessary calculations.

Changing the data

The charge for depreciation appears as an expense on the income statement, so it has an impact on the operating profit of the business. Once the values to be used for depreciating an asset have been established, it is not acceptable to change them because it might improve the financial position of the business. Only if it can be proved that the original estimates were very inaccurate would it be seen as acceptable to make any changes. This stops a firm 'window dressing' the financial statements by either overvaluing or undervaluing its fixed assets, or by increasing or decreasing expenses and therefore operating profits.

WORKED EXAMPLE

Company A

The finance director decides to extend the expected useful life of the £50,000 asset from five years to 10 years. The residual value remains the same at £5,000.

Calculate the new annual depreciation charge using the straight line method. How will this change affect the income statement of the business?

$$(£50,000 - £5,000) \div 10 = £45,000 \div 10 = £4,500 \text{ per year}$$

This will increase the book value of fixed assets each year, and reduce the expenses shown on the income statement, which will in turn increase the operating profit.

PRACTICE QUESTION

2   Company B has decided to review the depreciation for the £60,000 vehicle that it has just bought. It wants to extend the life of the asset by three years and reduce the residual value to £4,000.

Calculate the new annual depreciation figure using the straight line method. Also calculate the net book value at the end of three and five years. Compare these to your original calculations in practice question 1. How will this change affect the income statement of the business?

Reducing balance depreciation

This method deducts the same percentage of an asset's value every year: for example, 30%. This method is less straightforward, but it can be argued that it gives a more accurate representation of the loss in value of an asset, which is usually greater in the early years. The same information is required and a decision about the percentage figure to be used must be taken. As demonstrated above, the choices made will have an impact on the financial statements of the business.

The formula for calculating the depreciation for a given year is:

> Depreciation for a given year
>> = the historic cost or NBV (depending on the year) × the percentage expressed as a decimal

WORKED EXAMPLE

Company A

The finance director has recommended that the company uses the reducing balance method to depreciate the new machinery bought for **£50,000**. It is still estimated to have a **five-year** useful life and a residual value of **£5,000**. The annual depreciation will be **40%** of the net book value (NBV).

Note that depreciation will continue while the NBV is greater than the residual value. If it falls below this figure, then there is no further depreciation.

Calculate the depreciation for the machinery using the reducing balance method.

Year 1: £50,000 × 0.4 = **£20,000**   This reduces the NBV to £30,000.

Year 2: £30,000 × 0.4 = **£12,000**   This reduces the NBV to £18,000.

Year 3: £18,000 × 0.4 = **£7,200**   This reduces the NBV to £10,800.

Year 4: £10,800 × 0.4 = **£4,320**   This reduces the NBV to £6,480.

Year 5: £6, 480 × 0.4 = **£2,592**

In year 5 the NBV would fall to £3,888. This is lower than the residual value of £5,000 so this higher figure will be recorded in the company accounts. This is illustrated in the figure below.

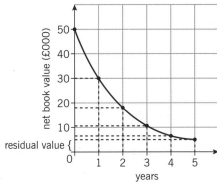

Graph showing the NBV of the machine over its five-year lifetime (reducing balance method)

PRACTICE QUESTION

3   The finance director of Company B has decided to investigate using the reducing balance method for depreciating the new vehicle because he believes that it is a more accurate approach for this type of fixed asset. The vehicle costs £60,000 and its estimated life is eight years, with a residual value of £4,000. The annual rate of depreciation chosen is 45%. Calculate the first year's depreciation and NBV using the reducing balance method.

Profitability ratios

These calculations enable the firm to analyse the relationship between revenue or capital spent and profits achieved.

Profit is the reward for risk and is the motive for those who invest in businesses. Shareholders will be interested to see whether profits have increased because this suggests that the return on their investment will also increase. Profitability ratios indicate the financial performance of a business more effectively than absolute profit levels.

More frequently, however, profit is expressed as a percentage of sales revenue or investment. This allows for better comparisons over time and with other businesses. Calculating profitability ratios also enables you to answer questions about the ability of a business to finance investment without needing to borrow money, or how efficiently the business is turning revenue into profit.

The following are the most common ratios that you will be expected to use to calculate profitability:

$$\text{Gross profit margin} = \frac{\text{gross profit}}{\text{revenue}} \times 100$$

$$\text{Operating profit margin} = \frac{\text{operating profit}}{\text{revenue}} \times 100$$

$$\text{Return on capital employed} = \frac{\text{operating profit}}{\text{total equity} + \text{non-current liabiities}} \times 100$$

or*

$$\text{Return on equity} = \frac{\text{operating profit}}{\text{total equity}} \times 100$$

*The formula used may depend on the examination board, so check the relevant specification for further details.

Gross profit margin

This calculates the difference between sales revenue and costs of sales. Costs of sales are direct costs because they are generated directly in the production of the good or service that is sold.

WORKED EXAMPLE

Company One plc
Calculate the gross profit margin for each of the two years shown opposite and make an initial analysis of the results.

$$\text{Gross profit margin} = \frac{\text{gross profit}}{\text{revenue}} \times 100$$

Year 1: (£4,523 ÷ £19,428) × 100 = 23.28%
Year 2: (£4,817 ÷ £23,009) × 100 = 20.94%

Extract from the income statement of Company One plc		
	Year 1 £m	Year 2 £m
Sales revenue	19,428	23,009
Cost of sales	14,905	18,192
Gross profit	**4,523**	**4,817**

The gross profit margin has fallen. This is because whilst sales revenue has increased by 18.43%, the cost of sales has increased by 22.05%.

Operating profit margin

This is profit after all expenses or indirect costs have been deducted, but before tax and interest are paid. It is useful as a measure of the efficiency of a business, because it relates to the costs of running the company rather than the costs of producing the good or service.

✓ WORKED EXAMPLE

UK Manufacturing plc

Use the information provided to calculate the operating profit margin and make an initial analysis as to whether or not the company has increased its efficiency.

Extract from the income statement of UK Manufacturing plc		
	Year 1 £m	Year 2 £m
Sales revenue	11,124	12,161
Gross profit	2,448	2,745
Expenses	1,262	1,372
Operating profit	**1,186**	**1,373**

$$\text{Operating profit margin} = \frac{\text{operating profit}}{\text{revenue}} \times 100$$

Year 1: $(£1,186 \div £11,124) \times 100 = 10.66\%$

Year 2: $(£1,373 \div £12,161) \times 100 = 11.29\%$

$$\text{Gross profit margin} = \frac{\text{gross profit}}{\text{revenue}} \times 100$$

Year 1: $(£2,448 \div £11,124) \times 100 = 22.01\%$

Year 2: $(£2,745 \div £12,161) \times 100 = 22.57\%$

The operating profit margin has increased by slightly less than the gross profit margin, indicating that the business is not operating more efficiently, even though the operating profit margin has risen. If the operating profit had improved significantly (e.g. by more than the gross profit margin), then it would suggest that the business is able to control expenses as a proportion of sales revenue. This would indicate that the business was operating more efficiently in year 2 than it was in year 1.

Return on capital employed/return on equity

This is a very important ratio because it calculates how efficiently the business is able to use the long-term capital invested in it. This can be used in comparisons over time, with other businesses and with other investment opportunities.

✓ WORKED EXAMPLE

The People's Theatre plc

Calculate the return on capital employed and the return on equity from the information given below and make an initial assessment of your findings.

Extract from the income statement of The People's Theatre plc	
	£
Revenue	38,235
Operating profit	17,105

Extract from the balance sheet of The People's Theatre plc	
	£
Non-current liabilities	45,800
Total equity	218,836

$$\text{Return on capital employed} = \frac{\text{operating profit}}{\text{total equity} + \text{non-current liabilities}} \times 100$$

$£17,105 \div (£218,836 + £45,800) \times 100 = 6.46\%$

$$\text{Return on equity} = \frac{\text{operating profit}}{\text{total equity}} \times 100$$

$(£17,105 \div £218,836) \times 100 = 7.82\%$

To assess the significance of the results of this calculation you would need information about previous years and other firms in the industry. However, if interest rates are very low, then a return of 6–7% may be acceptable.

WORKED EXAMPLE:
CALCULATING THE PROFITABILITY OF A BUSINESS

Global Company plc

The following are simplified accounts for a large manufacturing company.

Calculate the four profitability ratios for both years from the information given below.

Extracts from the balance sheet		
	Year 1 (£)	Year 2 (£)
Current assets	11,298,929	13,073,604
Of which inventories	1,459,394	1,422,373
Current liabilities	(10,589,293)	(10,686,214)
Non-current assets	17,763,108	17,275,683
Non-current liabilities	(7,872,007)	(8,732,630)
Total equity	10,600,737	10,930,443

Extract from the income statement		
	Year 1 (£)	Year 2 (£)
Sales revenue	20,529,570	18,950,973
Cost of goods sold	18,455,800	16,683,797
Gross profit	**2,073,770**	**2,267,176**
Expenses	2,534,781	2,119,660
Operating profit/loss	**(461,011)**	**147,516**

Gross profit margin: Year 1 $\dfrac{£2,073,770}{£20,529,570} \times 100 = \textbf{10.1\%}$ Year 2 $\dfrac{£2,267,176}{£18,950,973} \times 100 = \textbf{11.96\%}$

Operating profit margin: Year 1 $\dfrac{(£461,011)}{£20,529,570} \times 100 = \textbf{(2.25)\%}$ Year 2 $\dfrac{£147,516}{£18,950,973} \times 100 = \textbf{0.78\%}$

Return on capital employed: Year 1 $\dfrac{(£461,011)}{£10,600,737 + £7,872,007} \times 100 = \textbf{(2.50)\%}$

Year 2 $\dfrac{£147,516}{£10,930,443 + £8,732,630} \times 100 = \textbf{0.75\%}$

Return on equity: Year 1 $\dfrac{(£461,011)}{£10,600,737} \times 100 = \textbf{(4.35)\%}$ Year 2 $\dfrac{£147,516}{£10,930,443} \times 100 = \textbf{1.35\%}$

> **! REMEMBER: Changes have been made to the names of financial statements and items within them.**
>
> All recent textbooks and examination papers will use the latest financial vocabulary. However, if you look at older books and financial statements, alternative descriptions may be used. For example:
>
> turnover = revenue
>
> fixed assets = non-current assets
>
> long-term liabilities = non-current liabilities
>
> shareholder funds = total equity
>
> capital employed = total equity + non-current liabilities

PRACTICE QUESTIONS

1 Below are the simplified accounts for a national retail business, Stylish plc, that has been struggling in difficult trading conditions. Calculate the profitability of the organisation for the two years given using four accounting ratios and make an assessment of its performance.

Extracts from the balance sheet		
	Year 1 (£000)	Year 2 (£000)
Non-current assets	74,094	70,020
Current assets	66,862	59,600
Of which stocks	40,000	36,100
Current liabilities	(23,983)	(30,214)
Net current assets	42,879	29,386
Non-current liabilities	(33,000)	(30,000)
Net assets	83,973	69,406
Total equity	83,973	69,406

> **REMEMBER:** In accounting, rather than showing a negative number by using the minus sign, brackets are used.
>
> In the balance sheet, this usually applies to the current liabilities and the non-current liabilities. However, it is not unusual for the brackets to be missing. You have to remember that liabilities are **minuses** in the calculation of net assets.
>
> Examples:
> Current assets **minus** current liabilities
>
> Net current assets **minus** non-current liabilities

Extracts from the income statement		
	Year 1 (£000)	Year 2 (£000)
Revenue	200,000	198,000
Cost of sales	106,103	120,184
Gross profit	**93,897**	**77,816**
Expenses	53,441	58,362
Operating profit/loss	**40,456**	**19,454**

2 Below are the simplified accounts for Evans Outdoors. Calculate the operating profit and the return on capital employed for both sides of this business.

Extracts from the income statement		
	Clothing (£000)	Equipment (£000)
Gross profit	200	175
Expenses	128	50

Extracts from the balance sheet		
	Clothing (£000)	Equipment (£000)
Non-current liabilities	(680)	(340)
Total equity	2,900	2,900

Shareholder ratios

These calculations help shareholders to measure the value of the return they receive for their investment. For the majority of shareholders, this is the most important consideration.

Shareholders receive dividend payments made from distributed profits. They also gain if the market price of the shares increases beyond the amount that the shareholders paid for them.

To calculate the rewards to shareholders, use the following formulae:

$$\text{Dividend per share (DPS)} = \frac{\text{total dividends}}{\text{number of shares issued}}$$

or★

$$\text{Earnings per share (EPS)} = \frac{\text{net profit after tax}}{\text{number of shares}}$$

★Depending on the examination board. Check the relevant specification for details.

Dividends per share

WORKED EXAMPLE

Company A
The company has issued share capital of 30,500 £1 shares.

In year 1 the profits for distribution were £25,000.

In year 2 the profits for distribution were £20,000.

Calculate the dividends per share for each year and comment.

DPS = total dividends ÷ number of shares.

Year 1: £25,000 ÷ 30,500 = £0.82 (or 82p) per share

Year 2: £20,000 ÷ 30,500 = £0.66 (or 66p) per share

You would expect that the shareholders will be disappointed with the fall in dividend from 82p per share to 66p per share. However, it is very difficult to make a meaningful analysis without knowing how much the shareholders paid for the shares. Without this information you cannot calculate the true return they received on their investment.

Dividend yield

The **dividend yield ratio** gives a clearer indication of the return on the investment made by shareholders. It can be compared with returns from previous years and other investment opportunities such as dividend yields in other companies or interest paid on a savings account. Of course, shareholders would like to see the dividend yield increasing every year, but in difficult trading conditions this may not always be possible so the context should always be considered when analysing results. For this you can use the following formula:

$$\text{Dividend yield (DY)} = \frac{\text{dividend per share}}{\text{market price}} \times 100$$

WORKED EXAMPLE

Company A
The average market price for Company A shares on the London Stock Exchange over the two-year period was £6.50 (or 650p) per share. Calculate the dividend yield for each year and give an initial analysis.

DY = (dividend per share ÷ market price) × 100

Year 1: (£0.82 ÷ £6.50) × 100 = 12.62%

Year 2: (£0.66 ÷ £6.50) × 100 = 10.15%

The dividend yield decreased over the two–year period, which is not good news for shareholders. It means that the return on their investment has fallen. However, if interest rates were generally low, even 10.15% might be considered a good return.

PRACTICE QUESTION

1   From the information provided for Barder Computers plc, calculate the dividend per share and the dividend yield for each year and give an initial analysis.

	Year 1	Year 2
Number of issued ordinary £1 shares	£5.5m	£5.5m
Total dividends	£1.3m	£2.2m
Current assets	£5.2m	£6.1m
Current liabilities	£3.3m	£2.9m
Non-current liabilities	£14m	£16.5m
Reserves	£2.8m	£3.9m
One-off items	£0m	£2m
Operating profit	£8.2m	£11m
Market share price	£1.55	£2.22

Price/earnings ratio

The price/earnings ratio (PER) compares a company's current share price to its earnings per share. If it is high, the PER suggests that investors are expecting higher earnings growth in the future. It is always useful to compare the PER for one business with others in the industry or the market in general. This gives a better indication of future prospects. To calculate the PER you should use the following formula:

$$\text{Price/earnings ratio} = \frac{\text{market price}}{\text{earnings per share}}$$

> **REMEMBER:** Price/earnings ratios are normally calculated on the basis that all the profit made in the period is distributed, whether or not the profit is paid out to shareholders. This is important for analysis because it suggests that the calculations might be misleading if not all the profit is distributed.

WORKED EXAMPLE

Company A
The market price for shares over the past two years has averaged £6.50 per share. The earnings per share were £0.82 in year 1 and £0.66 in year 2. Calculate the PER for each year and give an initial comparison.

PER = market price ÷ earnings per share

Year 1: £6.50 ÷ £0.82 = £7.93   Year 2: £6.50 ÷ £0.66 = £9.85

This suggests that in year 1 shareholders were willing to pay £7.93 for a £1 return and in year 2 they were prepared to pay £9.85.

WORKED EXAMPLE: CALCULATING THE SHAREHOLDERS' REWARDS

Walshaw plc

This company, based in Barnsley, manufactures leather goods for a range of high street retailers. There is a French company very interested in buying a share of the business. Calculate the returns to shareholders for the past two years from the information given below. The market price for shares was £1.50 in year 1 and £1.25 in year 2.

Extracts from the balance sheet		
	Year 1 (£000)	Year 2 (£000)
Non-current assets	6,800	4,962
Current assets	10,582	9,231
Current liabilities	(7,287)	(6,580)
Non-current liabilities	(5,302)	(2,813)
Net assets	**4,793**	**4,800**
Issued ordinary share capital		
£1 nominal value	3,210	3,210
Reserves	1,583	1,590
Total equity	**4,793**	**4,800**

Extracts from the income statement		
	Year 1 (£000)	Year 2 (£000)
Revenue	27,203	21,558
Cost of sales	18,982	16,565
Gross profit	**8,221**	**4,993**
Expenses	4,084	3,678
Operating profit	**4,137**	**1,315**
Taxation and interest	281	202
Profit after tax	**3,856**	**1,113**

$$\text{Earnings per share (EPS)} = \frac{\text{net profit after tax}}{\text{number of shares}}$$

Year 1: £3,856 ÷ 3,210 = **£1.20**   Year 2: £1,113 ÷ 3,210 = **£0.35**

$$\text{Dividend yield (DY)} = \frac{\text{dividend per share}}{\text{market price}} \times 100$$

Year 1: (£1.20 ÷ £1.50) × 100 = **80%**   Year 2: (£0.35 ÷ £1.25) × 100 = **28%**

$$\text{Price/earnings ratio} = \frac{\text{market price}}{\text{earnings per share}}$$

Year 1: £1.50 ÷ £1.20 = **£1.25**   Year 2: £1.25 ÷ £0.35 = **£3.57**

PRACTICE QUESTION

2 From the information provided below for Gormally plc, calculate the earnings per share, the dividend yield and the price/earnings ratio. The market share price is £4.50.

Extracts from the income statement	
	£m
Revenue	610
Cost of sales	190
Gross profit	420
Expenses	240
Operating profit	180
Tax and interest payments	55
Profit after tax and interest	125

Extracts from the balance sheet	
	£m
Non-current assets	590
Current assets	81
Current liabilities	(80)
Net current assets	1
Non-current liabilities	(304)
Net assets	287
Ordinary share capital of £1 nominal value	260
Reserves	27
Total equity	287

REMEMBER: If you use the answer from one calculation as part of a further calculation, the **own figure rule (OFR)** will be applied.

For example, this will be the case when you use the dividend per share or earnings per share to calculate the dividend yield.

Liquidity ratios

Liquidity is a measure of a firm's ability to meet its current liabilities. The most common liquid assets held by firms are cash at the bank, inventories and receivables. Liquidity ratios compare the current assets to the current liabilities over a given period of time. The information required can be found in the balance sheet.

> **REMEMBER:** Just as other accounting language has changed, the balance sheet is now commonly known as the **statement of financial position**.

In this topic the following formulae will be used to calculate liquidity:

Current ratio = current assets ÷ current liabilities

Acid test ratio = (current assets − inventories) ÷ current liabilities

The **current ratio** indicates whether a business has enough liquid assets to pay its current liabilities. Generally you would expect the ratio to be between 1.5 : 1 and 2 : 1. Any less than 1.5 might suggest that the business lacks liquidity and any more than 2 : 1 means that money may be tied up unproductively: for example, in unsold inventories. However, as with other ratios, a judgement about liquidity based on the current ratio needs to be made in the context of the business. Some firms with very fast-moving inventories or those with seasonal products may have a current ratio that appears either too low or too high. Comparisons within the industry would be a helpful guide to what is satisfactory.

The **acid test ratio** is a more stringent test as it does not treat inventories as liquid assets. This is because inventories have not been sold and may instead become obsolete or be damaged. If the acid test ratio is less than 1 : 1 it indicates that the company cannot pay its current liabilities from its current assets minus inventories. Again, the significance of this needs to be given context. Changes over time or industry comparisons are useful guides as to whether or not it is a problem.

✓ WORKED EXAMPLE

A company to do business with?

The People's Theatre plc (TPT) is an independent company and a potential customer for a supplier of staging equipment. Use the information opposite and below to calculate the liquidity ratios for TPT and make observations.

Extracts from the income statement of TPT plc	
	£
Revenue	38,235
Operating profit	17,105

Extracts from the balance sheet of TPT plc		
	£	£
Non-current assets		
Land and buildings		258,000
Fixtures and fittings		19,850
Current assets		
Inventories	2,040	
Receivables	570	
Cash	11,562	
Current liabilities		
Payables	27,386	
Net current assets		(13,214)
Non-current liabilities		45,800
Net assets		218,836
Financed by:		
Issued ordinary share capital £1 nominal value		53,000
Reserves		165,836
Total equity		218,836

Current ratio = current assets ÷ current liabilities
= (£2,040 + £570 + £11,562) ÷ £27,386
= £14,172 ÷ £27,386
= **0.52 : 1**

$$\text{Acid test ratio} = \frac{(\text{current assets} - \text{inventories})}{\text{current liabilities}}$$
= (£14,172 − £2,040) ÷ £27,386
= £12,132 ÷ £27,386
= **0.44 : 1**

These figures show that the business is unable to meet its current liabilities. The current ratio figure is more of a problem because it includes all liquid assets. It shows that TPT only has enough liquidity to pay just over 50% of its current liabilities (i.e. the businesses to which it owes money). The acid test ratio shows the severity of the situation because if TPT takes out inventories, then it is only able to pay approximately 44% of its liabilities.

Note that some questions may require you to calculate some of the amounts on the income statement or balance sheet. The following formulae may be used:

Inventories = total current assets − receivables − cash

Net current assets = total current assets − total current liabilities

Net assets = non-current assets + net current assets − non-current liabilities

---

PRACTICE QUESTIONS

1 a Using the information given opposite and below, complete the statement by filling in the gaps marked by the letters (a) to (d).

b Calculate the current ratio and acid test ratios for both years. Show your workings.

Extracts from the income statement for Spindry plc (£m)		
	Year 1	Year 2
Sales revenue	105.00	102.00
Gross profit	89.00	86.00
Operating profit	62.00	66.00

> **! REMEMBER:** In this spread, **brackets have not always been used** to denote minus numbers. Remember that current and non-current liabilities are **deducted** from assets.

Extract from the balance sheet of Spindry plc (£m)		
	Year 1	Year 2
Non-current assets	102.00	99.50
Current assets		
Inventories	(a)	(b)
Receivables	37.50	14.60
Cash	22.30	25.90
Total current assets	**113.00**	**66.60**
Current liabilities		
Borrowings	0.00	4.10
Payables	47.20	26.20
Total current liabilities	**47.20**	**30.30**
Net current assets	(c)	36.30
Non-current liabilities	75.00	75.00
Net assets	**92.80**	(d)
Financed by:		
Share capital (£1 nominal value)	25.00	25.00
Reserves	67.80	35.80
Total equity	**92.80**	**60.80**

2 Use the information given below to calculate the liquidity ratios for this company.

Extracts from the balance sheet of Telemark plc	
	£m
Non-current assets	37,033
Current assets	12,465
Of which inventories	3,744
Current liabilities	(5,889)
Non-current liabilities	(14,483)
Net assets	16,661

> **! REMEMBER:** Some of the common financial vocabulary changes will feature in this topic.*
> For example:
> fixed assets = non-current assets
> stocks = inventories
> debtors = receivables
> creditors = payables
> long-term loans = non-current liabilities
> *Check with your examination board for clarification.

3 Look back over the other chapters on ratio analysis and calculate the current ratio and acid test (if possible) for Global Company plc (page 54), Stylish plc (page 55), Barder Computers plc (page 57), Walshaw plc (page 58) and Gormally plc (page 59).

Gearing

Calculating gearing

This involves measuring the percentage of a firm's capital that is financed by long-term loans (non-current liabilities). It is expressed as a percentage of total equity + non-current liabilities.

The gearing ratio is an important calculation because it shows how much of a firm's equity is made up of non-current liabilities, where interest payments are compulsory. When firms start to grow, they tend to raise finance from banks and other lending organisations which demand interest (often at high rates) as part of the repayment package. This reflects the risk involved in making the loan; the higher the risk, the higher the interest charged.

The alternative to loan capital is share capital, which does not have compulsory repayments. Shareholders do expect dividend payments as the reward for investing in the business, but the amount passed on to shareholders can be changed to reflect the level of profit achieved.

The following formulae can be used to calculate the gearing ratio:

$$\text{Gearing} = \frac{\text{non-current liabilities}}{\text{total equity} + \text{non-current liabilities}} \times 100$$

or

$$\text{Gearing} = \frac{\text{long-term liabilities}}{\text{capital employed}} \times 100$$

What is the right level of gearing?

This depends on the type of business. However, there are three classifications of gearing that might be helpful:

- A low-geared firm usually has less than 25% gearing. This suggests caution and low risk taking.
- A highly-geared firm usually has more than 50% gearing. This may be dangerous if the business has cash-flow problems and falling revenue. The business is also vulnerable to rises in interest rates.
- 25%–50% is usually 'normal' gearing for a 'mature' business, which is able to finance growth through both loan and share capital.

It is important to understand that having a highly-geared business might be a deliberate strategy of the owners or managers. For example, a high gearing ratio might suggest that business managers have a strategy of rapid growth. If this is successful, then higher profits resulting from this expansion will lead to much higher returns to shareholders. If the capital for expansion had been obtained from shareholders, the increased profit would need to be shared between more shareholders, so the rate of return they each receive would be less.

WORKED EXAMPLE

Company A

Calculate the gearing ratio from the information provided below.
Suggest whether the gearing is high, low or normal.

	Year 1 (£000)	Year 2 (£000)
Non-current liabilities	4,600	2,600
Total equity	5,000	6,450

Gearing = non-current liabilities ÷ (total equity + non-current liabilities) × 100

Year 1 gearing: £4,600 ÷ (£5,000 + £4,600) × 100 = 47.92%

Year 2 gearing: £2,600 ÷ (£6,450 + £2,600) × 100 = 28.73%

The gearing ratio has been lowered because the firm has reduced its non-current liabilities (for example, paying off a mortgage on some property) and increased the total equity (for example, increasing the reserves). Although the gearing has been lowered, Company A could be considered a normally geared business.

WORKED EXAMPLE

Alliance Boots

Calculate the gearing ratio for two years from the information provided below.

Extracts from the balance sheet for Alliance Boots		
	Year 1 (£m)	Year 2 (£m)
Non-current assets	13,228	13,563
Current assets	19,133	19,352
Of which inventories	2,030	1,782
Current liabilities	(5,712)	(4,561)
Net current assets	193	1,228
Non-current liabilities	(7,750)	(9,090)
Net assets	5,671	5,701
Financed by:		
Share capital	5,671	5,701
Total equity	5,671	5,701

> **REMEMBER:** The formula for calculating how figures have changed over time as a percentage is:
>
> (the change ÷ the original) × 100
>
> This is a very important formula to use when comparing data.

Gearing = non-current liabilities ÷ (total equity + non-current liabilities) × 100

Year 1: £7,750 ÷ (£5,671 + £7,750) × 100 = (£7,750 ÷ £13,421) × 100 = **57.75%**

Year 2: £9,090 ÷ (£5,701 + £9,090) × 100 = (£9,090 ÷ £14,791) × 100 = **61.46%**

Alliance Boots has increased its gearing ratio by increasing the level of non-current liabilities from £7,750m to £9,090m, a 17.29% increase, whilst only increasing share capital by £30m, a 0.53% increase. This explains the overall increase in the gearing ratio. Over the two-year period, Alliance Boots has continued to be a highly-geared company.

How can the gearing ratio be changed?

To reduce its gearing, a business needs to increase its total equity and/or reduce its non-current liabilities. For example, the company could:

- issue ordinary shares
- increase reserves by reducing dividend payments
- repay long-term loans such as mortgages.

Gearing is increased as a result of obtaining loan finance to pay for business expansion, which might lead to higher returns to shareholders. The business could take out additional long-term loans or sell debentures.

If a question asks you to 'recalculate the gearing ratio' after an increase in long-term loans to purchase premises or equipment, you must increase **both** the non-current liabilities figure and the capital employed figure by the amount of additional capital raised.

WORKED EXAMPLE

A business has a gearing ratio of 50% (capital employed = £1m and non-current liabilities = £0.5m). It increases loans by £0.25m to purchase assets of the same value.

Calculate the new gearing ratio.

$$\text{Gearing} = \frac{\text{long-term liabilities}}{\text{capital employed}} \times 100$$

The new gearing ratio becomes:

$$\frac{£0.75\,\text{m}}{£1.25\,\text{m}} \times 100 = 60\%$$

PRACTICE QUESTIONS

1 Calculate the gearing ratio for each year for Global Company plc and suggest whether the gearing is high, low or normal.

	Year 1	Year 2
Long-term liabilities	7,872,007	8,732,630
Capital employed	18,472,744	19,663,073

2 Use the gearing ratio to explain what has happened to the capital structure of Company B.

Extracts from the balance sheet of Company B		
	Year 1 £000	Year 2 £000
Non-current liabilities	2,381	5,230
Issued ordinary share capital	3,120	3,120
Reserves	589	1,070

STRETCH YOURSELF
OTHER CAPITAL STRUCTURE RATIOS

Comparing assets, liabilities and shareholder equity

Gearing is also known as **leverage** and there are three other ratios to consider.

a The **debt to equity ratio** compares the total interest-bearing debt of a business with the shareholder equity and indicates whether the company has been aggressive in financing its growth with debt. The following formula is used:

$$\text{Debt to equity ratio} = \frac{\text{total interest-bearing debt}}{\text{shareholders' equity}}$$

If the ratio is close to 1, and is different from the normal ratio in that industry, then the firm is normally believed to have been aggressive in financing its growth by debt.

Example: Company A

Total interest-bearing liabilities   £238m

Ordinary share capital   £312m

Calculate the ratio, and make an initial analysis of its significance in an industry where the usual ratio is 0.4 : 1.

b The **debt to assets ratio** again looks at how a company is financed by comparing total liabilities and total assets using the formula:

$$\text{Debt to assets ratio} = \frac{\text{total liabilities}}{\text{total assets}} \times 100$$

The calculation includes all current and non-current assets and liabilities. It is generally agreed that a low ratio (percentage) is preferred, as it suggests that the firm is a low risk.

Example: Company B

Non-current assets £37,033m; current assets £12,465m; current liabilities £5,889m; non-current liabilities £14,483m

Calculate the debt to assets ratio and comment on the level of risk that Company B presents.

c The **proprietary ratio** shows the contribution of shareholders to the total capital of a company. A high proprietary ratio indicates a strong financial position and good security for creditors. However, it may also suggest that the firm has not taken advantage of loan capital available and growth has been restricted because of this.

Two formulae are used, one including intangible assets and the other not. It is worth considering the importance of intangible assets for particular businesses when deciding which formula to use.

$$\text{Proprietary ratio 1:} \quad \frac{\text{shareholders' equity}}{\text{total assets}} \times 100$$

$$\text{Proprietary ratio 2:} \quad \frac{\text{shareholders' equity}}{\text{total assets} - \text{intangibles}} \times 100$$

Example: Company C

Total assets £950,000m; intangible assets £150,000m; shareholders' equity £440,000m

Calculate the proprietary ratios and offer an initial analysis of your findings.

Financial efficiency ratios

These ratios help to assess how effectively a business is managing its assets, including inventories (stocks) and receivables (debtors).

Asset turnover

This ratio measures how 'hard' the firm's assets are working to produce revenue for the business; the higher the ratio, the more efficiently the assets are being used. Asset turnover is useful for analysing the performance of a business over time and in comparison with other firms in the industry. It can also be used within a large organisation to compare different profit centres. The formula is:

$$\text{Asset turnover} = \frac{\text{revenue (sales)}}{\text{net assets}}$$

WORKED EXAMPLE

Global Company plc

Year 1: Revenue £20,529,570; net assets £10,600,737

Year 2: Revenue £18,950,973; net assets £10,930,443

The industry average asset turnover is 0.85.

Calculate the asset turnover for both years and make some initial observations.

Year 1: Asset turnover = revenue ÷ net assets
$$= £20,529,570 \div £10,600,737$$
$$= 1.94$$

Year 2: Asset turnover $= £18,950,973 \div £10,930,443$
$$= 1.73$$

If the industry average was 0.85, this would suggest that Global Company plc is using its capital efficiently, although the efficiency was greater in year 1 than in year 2, which may be worth investigating further.

Inventory turnover

This ratio calculates the number of times inventories are used and replaced over a year. As a general rule, the higher the inventory turnover, the more efficient the business because it is avoiding excess inventory holdings and minimising the opportunity cost of capital tied up in inventories of raw materials, semi-finished or finished goods. However, there are many factors that explain inventory turnover: for example, perishable goods need to be turned over very quickly for obvious reasons, whereas a jeweller may need to carry a large stock to show clients, and does not anticipate a high turnover. As with other ratios, it is important for one firm to compare its inventory turnover ratio with that of other firms in the industry. The inventory turnover formula is:

$$\text{Inventory turnover} = \frac{\text{cost of sales}}{\text{average inventories held}}$$

where average inventories held may be calculated as follows:

$$\frac{\text{opening inventories} + \text{closing inventories}}{2}$$

WORKED EXAMPLE

Mobile Phone Company (£m)

Year 1: Cost of sales £17,896; average inventories £365

Year 2: Cost of sales £13,446; average inventories £513

Calculate and comment on the inventory turnover.

Year 1: Inventory turnover = cost of sales ÷ average inventories
= £17,896 ÷ £365 = 49

Year 2: Inventory turnover = cost of sales ÷ average inventories
= £13,446 ÷ £513 = 26

Change in inventory turnover = (change ÷ original) × 100
= (49 − 26 = 23 ÷ original 49) × 100 = **47%**

This business should be concerned that its inventory turnover rate has dropped by 47%. This would be particularly significant in such a competitive market, where new models of mobile phone can make inventories obsolete very quickly.

Receivables (debtor) days

This ratio looks at how long it takes debtors to pay money that they owe. There are general principles to indicate whether or not a business is allowing its customers too much trade credit and has a problem with collecting debts. This ratio has a strong link to cash flow within a business and is therefore an indicator of financial health and good or poor financial management.

The formula for receivables days is:

Receivables days = (receivables ÷ sales revenue) × 365

WORKED EXAMPLE

Spindry plc (£m)

Year 1: Sales revenue £360.40; receivables £23.50

Year 2: Sales revenue £313.80; receivables £28.30

Calculate and comment on the receivables days.

Year 1: Receivables days = (£23.50 ÷ £360.40) × 365 days = 24 days

Year 2: Receivables days = (£28.30 ÷ £313.80) × 365 days = 33 days

The customers of this business pay in just over one month, which will certainly help to maintain a healthy cash flow. However, the fact that the repayment time has increased by nine days may be a cause for concern.

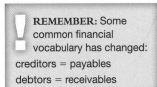

REMEMBER: Some common financial vocabulary has changed:
creditors = payables
debtors = receivables

Payables (creditor) days

This ratio calculates how long it takes a business to pay its suppliers the money they are owed. Suppliers are likely to look at this ratio to assess the reliability of a business before deciding to accept orders. As with receivables days, this ratio may indicate the financial health of a business.

The formula for payables days is:

Payables days = (payables ÷ cost of sales) × 365

WORKED EXAMPLE

Spindry plc (£m)

Year 1: Payables £36.20; cost of sales £135

Year 2: Payables £32.40; cost of sales £150.40

Calculate and comment on the payables days.

Year 1: Payables days = (£36.20 ÷ £135) × 365 = 98 days

Year 2: Payables days = (£32.40 ÷ £150.40) × 365 = 79 days

These calculations suggest that in year 1, Spindry took more than three months to pay suppliers. This has been reduced by 19 days, but it is still a very long time. This may cause cash-flow problems for suppliers; however, it certainly helps cash flow at Spindry.

PRACTICE QUESTIONS

1   Calculate the financial efficiency ratios of Company One plc using the information provided below. Make some initial observations about the position of the company in year 2 compared to the previous year.

Extracts from the income statement for Company One plc		
	Year 1 £m	Year 2 £m
Sales revenue	19,428	23,009
Cost of sales	14,905	18,192

Extracts from the balance sheet for Company One plc		
	Year 1 £m	Year 2 £m
Inventories (average)	2,069	1,782
Receivables	3,530	3,078
Payables	(4,603)	(4,172)

2   Use the extracts from the financial statements of UK Manufacturing plc to calculate the four financial efficiency ratios required to assess how effectively the business is managing its assets. Compare them with Company One plc.

Extracts from the income statement for UK Manufacturing plc		
	Year 1 £m	Year 2 £m
Revenue	11,124	12,161
Cost of sales	(8,676)	(9,416)
Gross profit	2,448	2,745
Operating profit	1,186	1,373

Extracts from the balance sheet of UK Maufacturing plc		
	Year 1 £m	Year 2 £m
Non-current assets	8,108	8,522
Current assets	8,315	9,593
Of which Inventories (av.)	2,561	2,726
Receivables	4,009	4,119
Current liabilities	(6,916)	(7,194)
Of which Payables	(2,356)	(2,571)
Non-current liabilities	(4,988)	(4,816)
Net assets	**4,519**	**6,105**
Financed by		
Share capital	374	374
Other capital including retained profits	4,145	5,731
	4,519	**6,105**

STRETCH YOURSELF

Use the information provided about UK Manufacturing plc to calculate as many ratios as you can remember. In each case, make an initial assessment of the implications of your results.

Look for more company results online − more than one business in the same sector if possible. This will enable you to make comparisons across industries as well as over time.

Simple payback and discounted payback period

Investment is an important part of managing a business as it concerns using current assets to produce future earnings for the business in the form of increased revenues or cash inflows. An understanding of the concept of opportunity cost is also relevant here: that is, the cost of a strategic decision in terms of the next best alternative sacrificed.

Investment appraisal is concerned with how a business decides whether or not an investment project is worthwhile. If there is more than one proposal, investment appraisal can be used to decide which is the most likely to benefit the firm. Investment appraisal involves a range of numerical techniques, some more precise than others. However, as with all such calculations, the reliability of the data used should always be considered.

Payback refers to the time taken for an investment project to recover the **initial cost** of the investment. It requires an estimate of the **useful life** of the project: that is, how many years they expect the investment to make a positive contribution to the business, and the expected inflows and outflows of cash during that life (the **net cash flow**).

Calculating payback

In order to calculate the payback period to the nearest month, it is necessary to calculate the cumulative cash flow of the project.

For example, a machine costs £10,000 and has an expected life of five years. The net cash flow at the end of year 1 is £6,000, so the amount outstanding on the cost of the machine is £4,000 (£10,000 − £6,000). At the end of the second year, the net cash inflow is £4,000, which means that the initial outlay of £10,000 has been paid back by the end of year 2. The cumulative cash flow of £10,000 (£6,000 + £4,000) = the initial outlay of £10,000.

If it is clear that the investment will be recovered during a particular year, the following formula is used:

$$\text{Payback} = \frac{\text{income required}}{\text{net cash flow for the year}} \times 12 \text{ (months)}$$

For example, if the machine bought for £10,000 gave a net cash flow of £4,000 in year 1 and £4,000 in year 2, then there would still be £2,000 outstanding at the end of the year. If the net cash flow for year 3 is £4,000, then the initial outlay will be paid back at some time during the third year. By applying the above formula, the payback period can be calculated to **the nearest month**:

$$\text{Payback} = (£2,000 \div £4,000) \times 12 = 6 \text{ months}$$

Therefore the payback period for this machine is two years six months.

A business will want the payback period to be a short as possible, and this type of appraisal is particularly useful when:

- a firm has cash–flow problems and liquidity is more important than profitability
- technology is changing rapidly and there is a danger that the investment may become obsolete, despite short-term cost and efficiency advantages
- there are several projects under consideration and the business has need of a simple screening process.

REMEMBER: You may prefer to give your answer as 2.5 years, for example, not bothering to multiply by 12. This is acceptable; however, it is not as clear. Using this method can also lead to mistakes:

Note: 0.25 = 3 months, not 4 months. This is a very common error, so beware!

When calculating to the month, you should round **up** to the next month, not round down to the previous month.

✔ WORKED EXAMPLE

Company A

The operations director has proposed that the firm buys a new machine for £50,000 that will benefit the business by increasing productivity and enabling an increase in potential output for the next **five years**. This is appropriate as the business has experienced increased demand for its products for the past 12 months. However, there are cash-flow problems because of trade credit terms offered to customers, and the board of directors wants the investment to pay for itself within **three years**.

Using the information provided below, calculate the payback time for the investment project, and make an initial assessment as to whether or not the board should agree to go ahead with the proposal.

Year	Net cash inflow (£)
0	(50,000)
1	25,000
2	20,000
3	15,000
4	10,000
5	5,000

It is unlikely that you will be given the cumulative cash flow, so you will have to create your own column in order to make the necessary calculations.

Year	Net cash flow (£)	Cumulative cash flow (£)
0	(50,000)	(50,000)
1	25,000	(25,000)
2	20,000	(5,000)
3	15,000	10,000
4	10,000	20,000
5	5,000	25,000

Notice that in year 0 the net cash flow is negative. This is because the machine has to be paid for at a cost of £50,000, which is a cash outflow, but it is not bringing in any revenue (cash inflow) yet because it has not started to be used. By the end of year 1, £25,000 has been recovered, leaving only another £25,000 in order for the investment to be completely paid back. By the end of year 2 this has been reduced to £5,000 as £20,000 was recouped during that year. This means that payback occurs during year 3. We now introduce the formula:

Payback = (income required ÷ net cash flow for the next year) × 12

The **income required is £5,000** as this is the amount outstanding from the original investment of £50,000 (£25,000 year 1 + £20,000 year 2 = £45,000 recouped).

The **net cash flow for year 3 is £15,000** according to the table above.

(£5,000 ÷ £15,000) × 12 = 4 (months)

This means that at the end of 4 months, another £5,000 will have been recouped by the company and the initial outlay of £50,000 will have been recovered in full.

The payback period for this investment is **2 years and 4 months**.

This meets the requirements of the board of directors and therefore on the basis of this calculation the proposal should be accepted.

PRACTICE QUESTION

1   A business has the opportunity to invest in new equipment which will
    cost £60,000 and last for five years. The estimated net cash flows
    are given below. The firm is keen that the initial outlay is recouped
    as quickly as possible because it has had cash-flow problems in the
    past and needs to show the bank that it will be able to pay back the
    £60,000 loan required to make the purchase. The bank would like the
    payback period to be no more than four years. Using the information
    provided, calculate the payback period and assess whether or not the
    business is likely to get the loan on the basis of your findings.

Year	Net cash flow (£000)
0	(60)
1	10
2	20
3	20
4	20
5	20

Years and months

Sometimes the payback calculation will fall exactly at the end of a year. This is
to be expected and does not indicate that your calculation was incorrect.

PRACTICE QUESTION

2   A retailer is interested in leasing a new store and has asked for
    information to assess the likely payback period for this investment
    opportunity. It is expected that the initial cost including new fixtures and
    fittings will be £12m for a five-year lease. Calculate the payback.

Year	Net cash flow (£m)
0	(12)
1	2
2	3
3	3
4	4
5	5

Net cash flow

Although a project will have an initial cost, it is likely that there will be
ongoing costs (cash outflows) associated with the project as well as revenues
(cash inflows) achieved. If you are only given the cash inflows and cash
outflows, you will have to calculate the net cash flow before you can proceed
with working out the payback. The net cash flow is calculated by subtracting
cash outflows from cash inflows.

✓ WORKED EXAMPLE

Company B

This business has the choice of two alternative investment projects, both costing **£10,000**, and both with an estimated life of **five years**. The firm operates in a high-technology industry where products and processes can become obsolete within five years, so the directors are looking for a short payback period. Use the information provided below to calculate which project is best on the basis of payback.

Year	Project A		Project B	
	Cash inflow (£000)	Cash outflow (£000)	Cash inflow (£000)	Cash outflow (£000)
0	0	10	0	10
1	6	3	7	6
2	8	3	8	5
3	8	5	9	4
4	8	6	10	6
5	7	4	11	5

In this question, you have to calculate the net cash flow (NCF) and then the cumulative cash flow (CCF) in order to calculate the payback period. This means adding four columns to the table given above.

Year	Project A (£000)				Project B (£000)			
	CI	CO	NCF	CCF	CI	CO	NCF	CCF
0	0	10	(10)	**(10)**	0	10	(10)	**(10)**
1	6	3	3	**(7)**	7	6	1	**(9)**
2	8	3	5	**(2)**	8	5	3	**(6)**
3	8	5	3	1	9	4	5	**(1)**
4	8	6	2	3	10	6	4	3
5	7	4	3	6	11	5	6	9

Project A: payback occurs in year 3.

Using the formula: (income required ÷ net cash flow for the next year) × 12

$(2 \div 3) \times 12 = 8$

Therefore payback is **two years and eight months**.

Project B: payback occurs during year 4.

$(1 \div 4) \times 12 = 3$

Therefore payback is **three years and three months**.

On the basis of this calculation only, the company should select Project A because it has the shorter payback period.

STRETCH YOURSELF

Burkinshaw plc

This manufacturing company has two investment opportunities, but cannot afford to finance both of them. It is going to use payback to decide on the best alternative to choose, even though the company does not have any liquidity problems. Use the information provided to calculate payback for both options. Make an initial assessment as to whether or not you agree that this is the best way to decide which investment project to choose.

> **REMEMBER:** If you have a decimal answer in your payback calculations (e.g. 2.35 months), always round up to the next month because the final payment will be made during the month in question.

Year	Project 1: £100m		Project 2: £80m	
	Cash inflows	Cash outflows	Cash inflows	Cash outflows
0	0	100	0	80
1	50	30	30	15
2	70	40	57	20
3	90	50	65	24
4	110	40	74	26
5	90	40	74	26

Discounted payback period (DPP)

The disadvantage of simple payback is that it ignores the **time value** of money. In order to solve this problem, DPP discounts the net cash flows of the project to calculate the present value of each net cash flow. Once this is completed, the process is the same as the simple method. As with the net present value method of investment appraisal (see pages 80–83), the discount rate will be given to you, and it will have an impact on the outcome of the calculation.

WORKED EXAMPLE

Company A

Another investment proposal is put to the company directors which involves buying a machine with a life span of five years at a cost of £75,000. The net cash flow is given in the table below along with the interest rate of 10% which is expected to apply throughout the five-year period.

Year	Net cash flow	10% discount rate
0	(75,000)	1
1	35,000	0.91
2	30,000	0.83
3	30,000	0.75
4	20,000	0.68
5	15,000	0.62

Calculate the discounted payback period.

You will have to add two more columns to calculate the cumulative discounted cash flows.

Year	Net cash flow	10% discount rate	Discounted cash flows	Cumulative discounted cash flows
0	(75,000)	1	(75,000)	(75,000)
1	35,000	0.91	31,850	(43,150)
2	30,000	0.83	24,900	(18,250)
3	30,000	0.75	22,500	4,250
4	20,000	0.68	13,600	17,850
5	15,000	0.62	9,300	27,150

As with previous examples, we can see that the payback period falls between years 2 and 3. By using the formula: (income required ÷ discounted cashflow for the year) × 12, we can narrow the payback period down to the nearest month.

$$\text{Income required} = £18,250$$
$$\text{Discounted cash flow for the year} = £22,500$$
$$(£18,250 ÷ £22,500) × 12 = 9.7$$
$$= 10 \text{ months}$$

Therefore the discounted payback period would be 2 years 10 months.

The discounted payback period is a more reliable method, but it still does not take into account any of the net cash flows achieved after the payback period is reached.

PRACTICE QUESTION

3 A manufacturing company is considering an investment opportunity which would involve buying new equipment at a cost of £60,000. It has an estimated life of five years and the rate of interest is expected to be 10% throughout the five-year period. From the information given below, calculate the discounted net cash flows, the cumulative discounted cash flows and the discounted payback period for the investment.

Year	Net cash flow	10% discount rate	Discounted cash flows	Cumulative discounted cash flows
0	(60,000)	1		
1	25,000	0.91		
2	20,000	0.83		
3	18,500	0.75		
4	15,000	0.68		
5	10,000	0.62		

Average (or accounting) rate of return

The problem with the payback method of investment appraisal (see pages 70–5) is that is does not take into account all cash flows beyond the payback period. This means that it does not measure the overall profitability of an investment project. The average or accounting rate of return (ARR) expresses the profits arising from an investment proposal as a percentage of the initial capital cost.

The formula used to calculate the ARR is:

$$\text{Average rate of return} = \frac{\text{net return (profit) per year}}{\text{capital outlay}} \times 100$$

As with payback, the biggest advantage of using this method of investment appraisal is its simplicity. It is expressed as a percentage return on investment, which makes it easy to understand, particularly when decision-makers are used to using ROCE (see page 53) as a measure of the profitability of the business.

At A Level, business examination board specifications assume that annual profit is the same as annual net cash flows from an investment project.

> **REMEMBER:** Alternatives to the above formula may include:
>
> Average annual profit ÷ cost of investment
>
> Average annual return ÷ initial investment
>
> Average annual profit ÷ initial cost
>
> Average annual profit ÷ average capital outlay
>
> Your examination board specification will make clear which formula to use.

The percentage return can be compared to other potential projects and to other investment opportunities available to the business.

WORKED EXAMPLE

Company A

The operations director has proposed that the firm buys a new machine for £50,000 that will benefit the business by increasing productivity and enabling an increase in potential output for the next five years. This is appropriate as the business has experienced increased demand for its products. The human resources director, however, has proposed that the company set up an online training facility for all staff to improve the skills of the workforce and reduce the number of complaints and faulty goods returned to the company. This will have an initial cost of £50,000 and is expected to run for three years. The company cannot afford both projects.

Use the information provided to calculate the ARR for each project. Decide which offers the best return to the business.

Years	Operations project (£000)	Human resources project (£000)
	Net cash flow	Net cash flow
0	(50)	(50)
1	25	25
2	20	20
3	15	15
4	10	0
5	5	0

$$\text{Average rate of return} = \frac{\text{net return (profit) per year}}{\text{capital outlay}} \times 100$$

Operations project:

$$\text{net return (profit) per year} = \text{total net cash flow} \div \text{number of years}$$
$$= £75{,}000 \div 5 \text{ years}$$
$$= £15{,}000$$
$$\text{ARR} = (£15{,}000 \div £50{,}000) \times 100 = 30\%$$

Human resources project:

$$\text{net return (profit) per year} = £60{,}000 \div 3 \text{ years}$$
$$= £20{,}000$$
$$\text{ARR} = (£20{,}000 \div £50{,}000) \times 100 = 40\%$$

Based on the ARR calculation alone, the best option for Company A is to select the proposal from the human resources director because it gives the highest average rate of return.

REMEMBER: If you are given the cash inflows and cash outflows, calculate the net cash flow by subtracting outflows from inflows.

PRACTICE QUESTION

1 Company D has the choice of two alternative investment projects, both costing £400,000, and both with an estimated life of five years. Using the information provided, calculate the average rate of return for both alternatives, then make a recommendation as to which project should be chosen based on your findings.

Year	Project	A	Project	B
	Cash inflow (£000)	Cash outflow (£000)	Cash inflow (£000)	Cash outflow (£000)
0	0	400	0	400
1	200	100	300	120
2	250	125	260	120
3	280	160	200	80
4	360	180	180	60
5	300	150	150	50

Drawbacks to using ARR

There are reasons why this method of investment appraisal is less than satisfactory, and should not be used as the only decision-making tool.

- It does not take into account the timings of the cash inflows. It may be that for the first few years there is no positive net cash flow, which may cause cash-flow problems for some businesses.
- The time value of money is not considered (see page 74).
- It is difficult to use the result effectively unless a comparison can be made. If there is only one project to be assessed, how do managers know if the return is good enough?

PRACTICE QUESTION

2 Company E is involved in heavy engineering, making plant and machinery for the construction industry. The research and development department has designed a new piece of equipment to meet the needs of clients operating in extreme weather conditions. The project will initially cost £500m and is expected to have a 10-year life. The net cash flows are given below.

Calculate the average rate of return on the project and make an initial assessment as to the viability of the investment for the company. You are advised to also calculate the payback period because the firm is keen to recoup the cost of the investment as soon as possible.

Year	Net cash flows (£m)
0	(500)
1	(300)
2	(100)
3	20
4	150
5	300
6	500
7	600
8	600
9	500
10	350

> **!** **REMEMBER:** When the formula you are using calculates a percentage, it is vital to include the percentage sign in your answer. You may drop a mark if you do not use the correct notation.

STRETCH YOURSELF

Scott Electronics plc

The company is concerned about loss of market share and believes that this is due to products being seen as 'old-fashioned' and unreliable. The board of directors has asked for proposals to move the business forward and two possible projects have been submitted for approval.

The operations director has proposed an investment project of £50 million which should have a life of four years. His plans involve updating existing machinery and introducing computer-aided manufacturing for the first time. This would increase productive potential and improve quality.

The marketing director has proposed investing £40 million in new product development. She has commissioned some market research and believes that, with the development of an updated range, the company could regain its position in the market and improve perceptions of the company and its products. The project would have a lifespan of five years.

Use the information below to calculate the ARR **and** payback periods for both proposals and make an initial judgement as to which project you would recommend.

Year	Operations project (£m)	Marketing project (£m)
	Net cash flow	Net cash flow
0	(50)	(40)
1	5	(20)
2	20	(10)
3	35	50
4	40	60
5	–	60

Net present value

Net present value (NPV) calculates the total 'present-day' value of an investment project by taking into account the 'time value' of money. It does this by discounting the future earnings back to their present value.

Money flows today have greater value than the same cash flows in future. One reason for this is that today's cash could be 'saved' at current interest rates and so would increase in monetary value. To reduce future cash inflows to today's values, discounting is used (the opposite of compound interest). The formula used is the inverse of the compound interest formula:

$$\text{Net present value} = \frac{1}{(1 + r)^n}$$

where n = numbers of years in the future and r = the rate of discount used.

Thus, at a discount rate of 10%, £1 in one year's time becomes worth:

$$\frac{1}{(1 + 0.1)} = £0.909.$$

Therefore 0.909 becomes the discount factor for discounting cash flows received in one year's time if the discount rate is 10%. The rate of discount used is usually the rate of return required from the investment project.

The main benefit of this method of investment appraisal is that it takes into account the fact that the value of money changes over time. This means that is gives greater importance to the earlier cash flows, as they are discounted less than later cash inflows. There is also a clear decision-making element to discounting: if the NPV is positive, the investment is worthwhile, whereas if it is negative, the investment should be rejected.

There are, however, disadvantages to this method. Firstly, it is more complicated than payback and average (accounting) rate of return. It may be difficult to explain to decision-makers. Secondly, the calculation depends on the discount rate chosen; if this is inappropriate, then profitable investment opportunities may be rejected. Thirdly, it is difficult to compare NPVs from different projects if the initial capital cost of them is very different.

At A Level the discount factors, based on the selected discount rate, will be given to you in the examination, so you do not have to calculate them.

WORKED EXAMPLE

Company A

The operations director has proposed that the firm buys a new machine for **£50,000** that will benefit the business by increasing productivity and enabling an increase in potential output for the next **five years**. This is appropriate as the business has experienced increased demand for its products for the past 12 months.

Use the information provided to calculate the net present value of the future net cash flows using the 10% discount factors provided.

Year	Net cash flow (£000)	Discount factors (10% discount rate)
0	(50)	1
1	25	0.91
2	20	0.83
3	15	0.75
4	10	0.68
5	5	0.62

To find the NPV of the net cash inflows, multiply the figure for each year by the relevant discount factor. The NPVs are added up to calculate the total return on the investment.

Year	Net cash flow (£000)	Discount factors	Discounted cash flows (£000)
0	(50)	1	(50)
1	25	0.91	22.75
2	20	0.83	16.60
3	15	0.75	11.25
4	10	0.68	6.80
5	5	0.62	3.10
			Net present value = 10.5

The NPV is positive, which implies that the project is worth investing in – at a discount rate of 10%.

The process

The calculations show the current (discounted) value of future cash inflows. You will have noticed above that the net present value is obtained by subtracting the initial cost of the project from the total discounted cash inflows. The initial investment cost is not discounted at all, as this payment is assumed to be made at the start of the project ('Year 0'), so the cost is the present-day value.

It is clear that the discount rate chosen is an important part of the decision, as is the cost of the investment. Changes to either of these elements can have a significant impact on the calculation and would impact on the final investment decision.

WORKED EXAMPLE

Company F

The marketing director has compiled the following figures to support a £100,000 investment proposal, lasting three years. He wants to use 8% discount factors, but the finance director thinks that 10% would be more realistic. How much difference would using different discount rates make to the viability of the project?

Year	Net cash flow (£000)	10% discount factors	8% discount factors
0	(100)	1	1
1	40	0.91	0.93
2	40	0.83	0.86
3	80	0.75	0.79

To find the answer, it is necessary to calculate two sets of NPV figures.

Year	Net cash flow (£000)	10% discount factor	Discounted cash flows @ 10% (£000)	8% discount factor	Discounted cash flows @ 8% (£000)
0	(100)	1	(100)	1	(100)
1	40	0.91	36.4	0.93	37.2
2	40	0.83	33.2	0.86	34.4
3	80	0.75	60.0	0.79	63.2
			Net present value = 29.6		**Net present value = 34.8**

Using the lower discount factors of 8% gives a higher total return on the investment. However, in both cases the NPV is positive, which suggests that the project should be supported.

PRACTICE QUESTIONS

1   Company C has the opportunity to invest in new equipment which will cost £60,000 and last for five years. The estimated net cash flows are given below. The discount rate used is 10%. Calculate the net present value of the investment proposal.

Year	Net cash flow (£000)	10% discount factors
0	(60)	1
1	10	0.91
2	20	0.83
3	20	0.75
4	20	0.68
5	20	0.62

2   The following net cash flows have been forecast by a manufacturer for the purchase of an automated machine:

Year	Net cash flows (£)
0	(15,000)
1	8,000
2	10,000
3	5,000
4	5,000

a   Calculate the simple payback period.

b   Discount all cash flows at a rate of discount of 10%.

c   Calculate the discounted payback period.

d   Calculate the net present value.

STRETCH YOURSELF

Internal rate of return (IRR)

This is the rate at which the NPV for the project would be zero.

This means that the present value of the cash inflows would equal the present value of its outflows and it would represent the 'break-even' NPV. It tells the investor how high the alternative market rate of return (e.g. the interest rate) would have to be before the investment would become an unattractive option.

The IRR is usually calculated as part of spreadsheet software, so there is no need for it to be calculated manually. The formula used is beyond the scope required of A Level students and first-year undergraduates.

However, the principle can be illustrated by using two discount factors to see which one gives the lowest NPV: that is, which one is closest to the IRR.

An investment project is going to cost £2,000, last four years, and has the following net cash flow figures:

Year	Net cash flow (£)	DCF @ 10%	DCF @ 7%
0	(2,000)	$1 \times (2,000) = (2,000)$	(2,000)
1	100	$0.91 \times 100 = 91$	$0.93 \times 100 = 93$
2	100	$0.83 \times 100 = 83$	$0.87 \times 100 = 87$
3	100	$0.75 \times 100 = 75$	$0.82 \times 100 = 82$
4	2,500	$0.68 \times 2,500 = 1,700$	$0.76 \times 2,500 = 1,900$
		NPV = (51)	NPV = 162

Clearly, the IRR is between 10% and 7%. Try 9% discount factors (values obtainable from the back of this book) and try to calculate NPV with this rate. You will find that you are nearly there!

Decision making: decision trees

This decision-making technique attempts to quantify the options available to a business and the choices to be made. It gives a visual representation of all the alternatives and introduces an element of risk assessment by estimating the probability of a financial outcome occurring.

There are four main components to each decision tree:

1 **Decision points**: these are represented by squares and indicate where a decision has to be made.

2 **Probable outcomes**: these are represented by a circle.

3 **Expected monetary values**: these are the financial profits of each decision, calculated by multiplying the value of the probable outcome by the probability. These might be estimates or based upon past results from similar decisions.

4 **Expected return**: this is the expected monetary value minus the costs (if any) of the decision.

WORKED EXAMPLE:
EXPECTED MONETARY VALUE

Company A

The operations director wants to update equipment to improve the quality of production. However, the marketing director wants to invest in a new promotional campaign.

Use the information provided below to calculate which option offers the best return to the business.

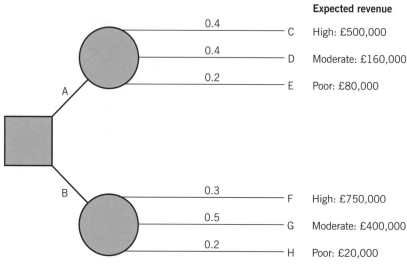

Expected revenue

C High: £500,000
D Moderate: £160,000
E Poor: £80,000

F High: £750,000
G Moderate: £400,000
H Poor: £20,000

A = new operations equipment

B = new promotional campaign

For each outcome the probability of success is expressed by a decimal value, such as 0.4. For each choice the probabilities should equal 1: for example, $0.4 + 0.4 + 0.2 = 1$.

There will always be the option of doing nothing, although this may not be shown in a given diagram.

For each option the expected revenue must be multiplied by the probability of success to calculate the expected monetary value (EMV).

Outcome C: £500,000 × 0.4 = £200,000
Outcome D: £160,000 × 0.4 = £64,000
Outcome E: £80,000 × 0.2 = £16,000
Total EMV for the operations equipment is:
£200,000 + £64,000 + £16,000 = **£280,000**.

Outcome F: £750,000 × 0.3 = £225,000
Outcome G: £400,000 × 0.5 = £200,000
Outcome H: £20,000 × 0.2 = £4,000
Total EMV for the promotional campaign is:
£225,000 + £200,000 + £4,000 = **£429,000**.

From these calculations we can see that the best outcome is the new promotional campaign having a high return, as this would give **£225,000**, although it has only a 30% chance of success. However, when all the expected monetary values are added together, it is clear that the promotional campaign has a much higher expected monetary value at **£429,000**, so this is the option that should be chosen. We have not yet included the cost of the two options.

Including initial costs

To make the calculations and comparisons more realistic, the costs of each project can be included. This gives the expected return of each investment project.

✓ WORKED EXAMPLE: EXPECTED RETURN

Company A
If the new equipment costs £50,000 and the promotional campaign costs £175,000, it is possible to calculate the expected return of each option.

A: EMV = £280,000 − £50,000 = **£230,000**

B: EMV = £429,000 − £175,000 = **£254,000**

The expected return on the promotional campaign is still higher than the return on the new operations equipment, despite the costs being much higher. Therefore, based on these calculations alone, the promotional project would be the preferred option.

Advantages of decision tree analysis

- The options and risks are set out clearly and logically.
- Decision trees show the amounts of money involved in each project and the likelihood of success for each possible outcome.
- By constructing a decision tree, courses of action not previously considered can be revealed: for example, to do nothing.
- Showing options in a visual form can help to explain alternatives to non-specialists.

Drawbacks to using decision tree analysis

Despite the advantages outlined above, this type of analysis should be used with caution.

- The probabilities are estimates and could therefore be inaccurate, out of date or liable to change.
- The data used are quantitative and do not include any consideration of the qualitative element of decision making.
- It would be quite easy to present biased data to favour one of the options.

PRACTICE QUESTION

1   Company B is faced with four possible options: to consolidate the existing range; to introduce a new model at a cost of £200,000; to redesign and relaunch an existing product at a cost of £90,000; or to increase promotion on the existing product range at a cost of £20,000. The probability factors shown in the diagram below are the result of extensive research among current and past Company B customers.

Using the decision tree, advise the business how it should proceed.

A = New model
B = Consolidate existing range
C = Redesign and relaunch existing product
D = Increase promotion of existing products

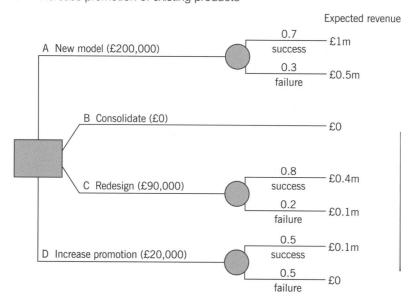

REMEMBER: You will see variations in the terms used for decision trees. Do not be put off by an unfamiliar word – the process will be the same. Work from right to left to find the final EMV or expected return.

STRETCH YOURSELF

Hairdressers' decision

Gavin and Bella are qualified hairdressers who own their own salon. It is popular and profitable. They only employ two junior assistants, as they take pride in the fact that they deal with each customer themselves. This close personal service has worked well up until now. Gavin and Bella's daughter is planning to marry in India next April and this is a very busy period for the salon. Gavin and Bella both want to go to the wedding, but they realise that this will mean being absent from the business for two weeks. They have considered three alternative options:

- Close the salon for the entire period and give up the £6,000 profit it would have made over the two weeks.
- Employ a professional hairstylist to take over the running of the salon for two weeks. She would cost £2,000 to employ but Gavin believes that there is a 50% chance of her making a £3,000 profit and a 50% chance of a £4,000 profit (before her salary is subtracted).
- Ask two friends to manage the business in their absence. They would receive a payment of £500 each. They know most of the customers personally, but their experience of hairdressing is limited to taking a course at college several years ago. Gavin believes that the two relatives have a 60% chance of making a profit (before payment) of £3,000 and a 40% chance of making £5,000.

a Draw a decision tree of the options available to Gavin and Bella and add to it the probabilities and possible returns of each option.

b Calculate the expected returns of the three options.

c Explain which option is best on financial grounds alone.

d Evaluate the other factors that you would advise Gavin and Bella to consider before taking the final decision.

Absenteeism, labour turnover and labour productivity

Businesses of every size want to ensure that they operate efficiently and profitably. This includes making sure that the workforce is a benefit rather than a cost to the firm.

Absenteeism

Absenteeism refers to employees who miss work without permission or a legitimate reason. It is a concern because it represents a cost to the firm. For example, output may fall, temporary staff may have to be employed at extra cost, and customer appointments may have to be cancelled, causing loss of business, loss of revenue and ultimately less profit. The same concerns arise in respect of employees who are late for work.

It is important for business managers to try to understand the causes of absenteeism and lateness and to look for possible ways to make improvements. However, information about absenteeism and lateness is useful for identifying trends and setting targets for improvement.

The formula used to calculate the level of absenteeism for a given time period is:

$$\text{Level of absenteeism} = \frac{\text{number of unauthorised absences}}{\text{total number of days worked by the workforce}} \times 100$$

WORKED EXAMPLE

Blackcountry Blades Ltd

This medium-sized manufacturing company makes wiper blades for the automotive industry from a factory in Walsall. There are 65 employees, who work an average of 282 days per year. The managing director, Robert Brewer, is concerned because he believes that there has been an increase in absenteeism over the last two years, and he has asked the HR manager for further details. The information gathered is given below.

	Two years ago	Last year	This year
Number of employees	75	70	65
Total number of days worked	21,150	19,750	18,330
Days lost to unauthorised absence	810	1,260	1,259

Calculate the absenteeism rate to see whether the managing director's fears are well founded.

Two years ago: $(810 \div 21{,}150) \times 100 = 3.83\%$

Last year: $(1{,}260 \div 19{,}750) \times 100 = 6.38\%$

This year: $(1{,}259 \div 18{,}330) \times 100 = 6.87\%$

Clearly there has been a significant increase in the rate of absenteeism, which needs further investigation.

PRACTICE QUESTION

1 In 2002 Catherine Booth and her friend Alice Foster set up their business, Catalice Home Care, looking after people's houses if they were away for at least one month. Their main clients were families who had holiday homes abroad, or worked for part of the year in another country, and the business has grown year on year. As a result Catalice was able to increase the workforce by 10% last year. In the early years, they found it very difficult to cope with high levels of absenteeism amongst their cleaning and maintenance staff, often reaching 20% per year. However since the recession, Catherine and Alice thought things had improved significantly, with rates falling to below 5%. However, last year 11,136 days were lost, which was the same number as two years ago. Has the trend changed, and should they be concerned?

On average, each member of staff works for 280 days per year. Complete the table below to calculate the absenteeism rate for the company for both years.

Catalice Home Care		
	Two years ago	**Last year**
Number of employees	580	
Total days worked (number of employees × 280)		
Days lost to unauthorised absence		
Absenteeism rate (%)		

Labour turnover

Labour turnover is concerned with the number of employees who leave a business per year. In fact, labour turnover includes people who join the business, but for the purposes of this calculation it is the leavers who are important. It is generally agreed that the higher the labour turnover, the more of a problem it is for the business, although rates will vary between industries. A high rate is likely to increase business costs. For example:

- recruitment and selection costs will rise
- productivity and output may fall
- customer service may suffer.

The formula that is used to calculate labour turnover is:

$$\text{Labour turnover} = \frac{\text{number of leavers that year}}{\text{average number of employees that year}} \times 100$$

As with many of these calculations, their value is as a tool for comparison over time and with other firms in the industry. The figures can also be used to see trends and set targets for improvement.

WORKED EXAMPLE

Blackcountry Blades

As well as dealing with the high absenteeism rate, Robert Brewer wants to maintain a labour turnover rate below the industry average of 22%. There are several competitors in the market that manufacture in the West Midlands and his skilled workers are sometimes tempted to go to rivals by offers of better pay and longer holidays. Robert believes that, although he has a core of loyal employees who have been with the company for many years, recent recruits are less likely to stay, despite all his attempts to use Herzberg's motivational techniques.

Use the data below to calculate the labour turnover for the last three years. Make an initial assessment of the situation in comparison to the industry average.

	Year 1	Year 2	Year 3 (last year)
Number of employees leaving	68	80	70
Workforce	340	320	280
Labour turnover rate	?	?	?

Year 1: $(68 \div 340) \times 100 = 20\%$

Year 2: $(80 \div 320) \times 100 = 25\%$

Year 3: $(70 \div 280) \times 100 = 25\%$

This suggests that, although the company's labour turnover rate is above the industry average of 22%, the situation appears to have stabilised. He should continue to explore reasons for the high labour turnover and possible solutions.

PRACTICE QUESTION

2   The owners of Catalice Home Care, Alice and Catherine, are looking into setting targets to reduce labour turnover and have called a meeting with all the regional managers to discuss strategies. They are particularly concerned about the South East region, which seems to have a much higher labour turnover than everywhere else. Calculate the labour turnover rates for the company as a whole and for the South East region to establish the extent of the problem.

Company data	National	South East
Number leaving this year	65	12
Current workforce	650	40
Labour turnover	?	?

Labour productivity

This calculates the output achieved per employee, and is often used as a measure of a firm's efficiency. Labour costs are an important part of total costs, so if the workforce becomes more productive, then costs per unit will fall. Labour costs are calculated using the following formula:

$$\text{Labour costs} = \frac{\text{total direct labour costs}}{\text{total units produced}}$$

However, there are problems associated with this approach:

- Should every member of the workforce be included in the calculation, or just direct labour?
- Is the productivity calculated for total output or by product?
- What if the business is a provider of services rather than a manufacturing company?

The formula used for calculating labour productivity is:

$$\text{Labour productivity} = \frac{\text{output per time period}}{\text{number of employees}}$$

 WORKED EXAMPLE

Potter's Brogues

Craig Potter employs four highly skilled shoemakers at his Nottingham branch and between them they can make an average of 136 pairs of shoes per month. Each shoemaker earns £25,000 per year. Calculate the annual rate of labour productivity and the unit labour costs for each pair of shoes.

136 × 12 = 1,632 pairs of shoes per year

1,632 ÷ 4 = **408 pairs of shoes per worker per year**

Total labour costs = £25,000 × 4 = £100,000

Unit labour costs:

total labour costs ÷ total units produced = £100,000 ÷ 1,632 = **£61.27**

Craig organises two weeks of intensive training with a visiting Italian shoe manufacturer, and the output per month increases to 148 pairs of shoes. Calculate the change in productivity and the new unit labour costs as a result of the training.

148 × 12 = 1,776 pairs of shoes per year

1,776 ÷ 4 = **444 pairs of shoes per worker per year**

The change in productivity is calculated using the formula:
(change ÷ original) × 100

The change in productivity = 444 − 408 = 36. The original productivity = 408, therefore (36 ÷ 408) × 100 = **8.8% increase**.

Unit labour costs: £100,000 ÷ 1,776 = **£56.31**

The change in unit labour costs: (£4.96 ÷ £61.27) × 100 = **8.1% reduction**

PRACTICE QUESTIONS

3   Demand for Craig's handmade shoes is growing, particularly from international customers, so he has increased his workforce (as given in the worked example above) by 50%. He has also bought each worker a new machine which speeds up production by 75% per month. Calculate the **change** in the annual labour productivity rate that Craig should expect.

4   An engineering firm, employing 10 workers who are each paid £1,670 per month, produces 1,000 components per month. As a result of buying new machinery, the same output can be produced by eight employees, so two workers are made redundant. Calculate the old and new productivity per worker per month and the old and new unit labour costs. Calculate the change in both the productivity and unit labour costs.

Critical path analysis (network analysis)

This is a technique which allows a business to manage complex projects. It is associated with lean production and also enables firms to estimate the minimum amount of time that a project will take.

In order to construct a critical path or network diagram, it is necessary to:
- set out all the activities involved in the completion of the project from beginning to end, including equipment required
- estimate accurately the length of time each activity involved in the project will take
- decide the order in which the activities have to be undertaken
- indicate which activities cannot start until other activities have been completed
- show which activities can be completed simultaneously – for example, those that do not depend on each other.

It is also important to account for delivery times, the availability of labour, and extraordinary considerations such as the weather.

WORKED EXAMPLE

Daisy's Dog Nursery

Daisy Priest has decided to produce an advertising campaign to promote her business, with the help of her sister, Rose. She is going to make a short video to show online and a poster for local newspapers and shop windows. She wants to show her campaign to her friend Sophia Kelly, who is studying marketing at university. The table below shows the estimated length of time each task will take and the order they need to be completed in.

Tasks	Order	Time (hours)
A	Plan the campaign: this must be done first	4
B	Make the video: can start when A is complete	8
C	Make the poster on the computer: can start when A is complete	5
D	Test market the video on a social networking site	24
E	Test market the poster on a social networking site	24
F	Present the campaign to Sophia	4
G	Launch the campaign, including distributing the poster	4

The total estimated time to complete the project is 71 hours, but Daisy knows that if they split up the work it can be done quicker. The question is: what is the shortest amount of time the project will take?

Construct a critical path diagram for Daisy and identify the critical path and total float.

The start and finish of each task is indicated by a circle called a node.

Each node is divided into three sections.
- The left-hand semicircle is the node number.
- This helps with tracking tasks through the network. The top-right quarter is the earliest start time (EST) and shows the earliest time that the next task can be started.
- The number in the bottom-right quarter is the latest finish time (LFT). This indicates the latest time that the previous task can be finished without delaying the start of the next task.

Arrows show the order in which the tasks are completed. The letter assigned to each task is shown above the arrow and the estimated completion time is shown below the arrow.

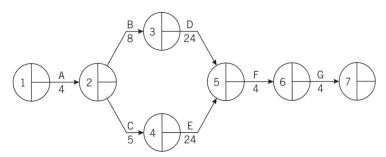

The diagram above shows us the complete network for all of the tasks in Daisy's campaign, with the tasks, node numbers and estimated completion times included. As you can see, the path splits in two and joins back up again. This is because activities B and C are independent from each other so can be undertaken simultaneously. D follows B and E follows C. F depends on both D and E so there is a single node from which F begins.

You can now add the ESTs and the LFTs, which will reveal the **critical path**.

To begin the process, start at the left-hand side of the diagram.

Earliest start time
The earliest **task A** can start is hour 0, so this goes in the top-right quadrant of node 1.

Tasks B and **C** cannot start until task A is finished, and this takes four hours. The EST for node 2 is therefore four hours (0 + 4) and this is indicated in the top-right quadrant. **Task D** cannot start until tasks A and B are complete, which is 12 hours (0 + 4 + 8), shown in node 3. **Task E** cannot start until tasks A and C are complete, which is 9 hours (0 + 4 + 5) and this is shown in node 4.

Task F is slightly more complicated because it requires tasks D and E to be complete. **Task D** takes 24 hours and so the EST for F is 36 hours (0 + 4 + 8 + 24), which is shown in node 5. This option is selected over route A → C → E → F which takes 33 hours (0 + 4 + 5 + 24) because you must look for the **longest route**. It is the work that takes the longest that must be completed on time if the campaign is to be launched on the date Daisy expects. Any delays along the critical path will push the completion date back and may spoil the effectiveness of the whole project.

Task F takes four hours so the EST in node 6 is 40 hours (0 + 4 + 8 + 24 + 4) and the EST for node 7 is 44 hours (0 + 4 + 8 + 24 + 4 + 4), which is the time it will take to complete the whole campaign based on Daisy's estimates.

Latest finish time
The next step is to identify the LFT for each task which does not increase the length of time that the project will take. To do this work back from right to left, starting at node 7. **Task G** must be finished by the 44th hour, so this goes in the bottom-right quadrant of node 7. To calculate the LFT for **task F** deduct the time it takes to complete task F from the previous LFT
(44 − 4 = 40). This figure goes in the bottom-left quadrant of node 6. You can establish the LFTs for **tasks D** and **E** in the same way (44 − 4 − 4 = 36), placing this in node 5.

Following the lower pathway, the LFT for **task C** is calculated to be 12 (44 − 4 − 4 − 24) and is shown in node 4. If the time for task C were then deducted, the LFT would be seven hours. However, following the other pathway we find that the LFT in node 3 is 12 (44 − 4 − 4 − 24) and if **task B** is then deducted, the LFT in node 2 would be four hours. In this case you are looking for the smallest number because this shows when the previous task must be finished by, so that the next task can begin. The smaller the number, the earlier in the process the task must be finished. This, therefore, is the number shown in node 2. The LFT in node 1 is 0 (44 − 4 − 4 − 24 − 8 − 4).

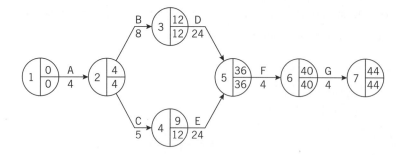

The critical path

This is revealed once all the nodes are complete as the route where the ESTs and the LFTs are the same. This means that there can be no delays between finishing one task and starting the next. This route, A → B → D → F → G, is critical because it takes the longest time and any delay will mean that the whole project cannot be finished in the 44 hours that Daisy estimated.

The total float

This is the identification of 'spare' time, which means that one or more tasks can be delayed and the process will still take the estimated completion time. To find the total float you need to look at the tasks that are not part of the critical path. In this case, subtract the EST (left-hand node) and the duration from the LFT (right–hand node) for task C.

The total float = the LFT of 12 (node 4) − 5 − 4 (node 2) = 3 hours. You could have completed the same process for task E and achieved the same result (36 − 24 − 9) = 3 hours. Be aware, though, that the 'spare' three hours can only be used once; if there are delays with task C of three hours, task E must be completed on time.

PRACTICE QUESTION

1 Demand for handmade trailers keeps increasing and Richard Thompson, the owner of the business, has decided to buy a piece of machinery which will speed up the production process without threatening the 'handmade' unique selling point of the business. With the help of the machine supplier, he has drawn up a network diagram (see below), showing the number of days each stage in the purchase and installation of the machine will take.

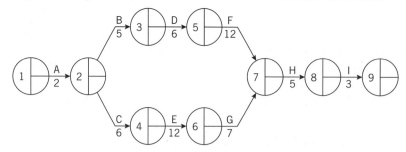

    a Calculate the EST for each task.
    b Calculate the LFT for each task and the minimum time it will take for the project to be completed.
    c Identify the critical path.
    d Identify the amount of total float.

Advantages of critical path analysis

- It provides a visual image of the stages in the completion of a project. This may help decision-makers to identify more easily the solutions to any problems that arise.
- It encourages forward planning and forces managers to consider all aspects of a project.
- It can improve efficiency. Orders relating to the project can be received 'just in time' and it can be an important part of lean production techniques.
- Cash flow can be controlled and improved because purchases can be made to coincide accurately with when they are required.

Disadvantages of critical path analysis

- It does not guarantee success. The times are often estimates and rely on a motivated workforce to ensure completion to the deadline.
- Complex projects might be difficult to put into a network diagram, although the development of computer software programs has assisted with this problem.
- It relies on the accuracy of the data used. These may not be reliable, especially if this is a new project or experience for the business.

STRETCH YOURSELF

Daisy's Dog Nursery

Refer to the worked example on pages 93–4. Daisy and Rose have begun preparing the promotional campaign, but they forgot to include printing time in the poster task. This will increase the number of hours to 10. How will this impact on the minimum time needed to complete the project, the critical path and total float?

Handmade Trailers

Refer to practice question 1. The project to install the new piece of machinery is about to start when Richard receives information from his supplier that the equipment has been delayed on its journey from Germany due to bad weather. This will increase task B by three days. He also hears that his senior electrician is on holiday when he is required to complete task G and this will slow things down by a further two days.

What impact will these changes have on the completion of the project?

Stock control

This is a method of stock management which aims to ensure that a business is holding the appropriate amount of raw materials and components, or finished goods, to meet demand.

The simplest form of stock control uses a graphical representation of a system based on the following features:

- a regular flow of sales during the time period (e.g. one month)
- a maximum stock level: the amount of stock the firm may wish to, or be able to, hold in order to meet demand in the given time period
- a minimum stock level: the smallest amount of stock that the business would wish to hold
- buffer stock: the difference between the minimum stock level and zero stock. This is held as a contingency in case of unexpected demand or other 'shock' situations
- the reorder level: the point which triggers a new stock order
- the reorder quantity: this is a constant amount in the simple model used here
- lead time: the time between stock being reordered and it arriving.

WORKED EXAMPLE

Potter's handmade shoes

As his business continues to flourish and grow, Craig Potter has introduced a stock control system to make sure that he has enough raw materials and components available to meet demand from his customers. The diagram below shows his stock control chart for shoelaces.

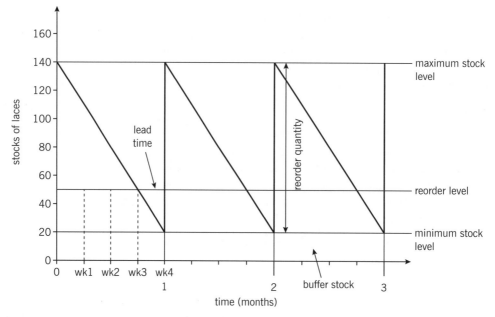

The chart shows that Craig has identified the **maximum stock level** as being 140 pairs of laces. This means that he thinks that he will never need more than that quantity of laces.

Craig has also decided that the lowest level of stock he should hold, known as the **minimum stock level**, is 20 pairs of laces. This gives him a **buffer stock** of 20 pairs in case there are delays in delivery.

120 pairs of laces are used every month and an order takes one week to be delivered. This is called the **lead time**. If we assume that there are four weeks in a month, we can calculate the level at which stock needs to be reordered:

$120 \div 4 = 30$, so we can assume that stock goes down at a rate of 30 pairs per week. Therefore:

Stock at week 1 = 140 − 30 = 110

Stock at week 2 = 110 − 30 = 80

Stock at week 3 = 80 − 30 = 50

Stock at week 3 = 50 − 30 = 20 (minimum stock level)

This means that in order to ensure there is enough stock to sell to customers, once stocks reach 50 pairs of laces, an order for 120 laces should be placed. This is known as the **reorder level**.

As you can see on the chart, the reorder level intersects the stock line at week 3 in each month.

 PRACTICE QUESTION

1 Richard Thompson of Handmade Trailers has used the same suppliers since starting his business making bespoke trailers. In looking for extra space to use for an increase in demand, he realises that he is carrying an excess stock of bolts used to secure the trailers. His supplier is in China and the stock control system that Richard agreed with the supplier is shown below.

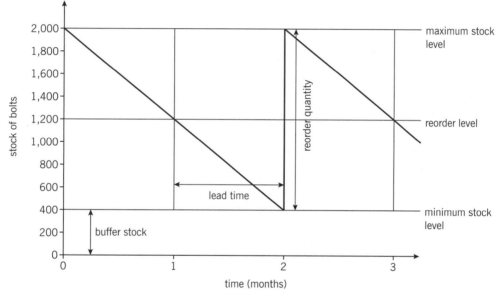

At a recent trade fair, Richard spoke to a bolt manufacturer from Birmingham and discussed the following terms.

The Birmingham firm can offer a lead time of one week to supply up to 1,000 bolts per month. Richard knows that he only uses about 800 bolts per month and is tempted by the quality of the product on offer and the lead time. This will free up storage space which could be put to better use. Ideally he would like to maintain a buffer stock of 200 bolts in case he gets a surprise order or a surge in demand, and would prefer a maximum stock level of 1,000 bolts.

a Use the chart above to identify the following:

   i the current maximum, minimum and buffer stock levels

   ii the reorder level, reorder quantity and lead time.

b Draw a graph to illustrate the new stock control chart Richard could use if he changed supplier. Make an initial analysis of whether or not he should change. Include details of the six features of the new chart as listed above.

The economic order quantity

STRETCH YOURSELF

The economic order quantity (EOQ)

This is a method of calculating the size of order for raw-material components or stocks which minimises total costs.

Total costs are made up of two elements:

- Acquisition costs such as negotiating with suppliers and administration. These costs fall as order quantities increase due to economies of scale.
- Holding costs such as rent for space and insurance, which increase with the order size.

The EOQ occurs where these two costs are the same.

It can be calculated using the following formula:

$$Q = \sqrt{2CA \div HP}$$

where: Q = EOQ; C = acquisition cost per order; A is the total number of units used each month; H is the holding cost as a percentage of the average stock value; and P is the price of each unit.

For example, a business uses 5,000 components per year, each costing £10. The holding cost of the components is 5% of the average stock value and the acquisition costs are £8.

$$Q = \sqrt{\frac{2 \times 8 \times 5{,}000}{0.05 \times 10}} = \sqrt{\frac{80{,}000}{0.5}} = \sqrt{160{,}000} = 400 \text{ components}$$

This shows that to minimise costs the company should buy 400 components at a time.

Richard Thompson at Handmade Trailers uses 9,600 bolts per year. Each bolt costs £2.50. The holding costs of the bolts is 10% of the average stock value and the acquisition costs are £2 per bolt.

What is the EOQ that minimises Richard's total costs?

List of formulae

Marketing

Market size by value $=$ total volume of sales $\times$ average market price

or

Market size by value $= \dfrac{\text{business sales}}{\text{\% market share}} \times 100$

Market share $= \dfrac{\text{value of firm A's sales}}{\text{value of total market sales}} \times 100$

Market growth $= \dfrac{\text{change in size of market by value}}{\text{original size of market by value}} \times 100$

Price elasticity of demand $= \dfrac{\text{percentage change in quantity demanded}}{\text{percentage change in price}}$ *or* $\dfrac{\%\Delta Qd}{\%\Delta P}$

Income elasticity of demand $= \dfrac{\text{percentage change in quantity demanded of good A}}{\text{percentage change in consumers' real income}}$

Accounting and finance

Total revenue $=$ average selling price $\times$ quantity sold

Total costs $=$ total fixed costs $+$ total variable costs

Average (unit) costs $=$ total costs $\div$ number of items produced

Profit $=$ total revenue $-$ total costs

Profit margin $=$ selling price $-$ unit cost

Budgeting

Operating profit $=$ gross profit $-$ expenses

Variance $=$ actual figures $-$ budgeted figures

Break-even analysis

Unit contribution $=$ selling price $-$ variable cost per unit

Total contribution $=$ total revenue $-$ total variable costs

Total contribution $=$ (selling price $-$ unit direct costs) $\times$ output or units sold

Change in contribution $=$ contribution in time period 2 $-$ contribution in time period 1

or

Change in contribution $=$ additional revenue $-$ additional variable costs

Break-even output $= \dfrac{\text{fixed costs of the business}}{\text{contribution per unit}}$

Financial accounts

Annual depreciation charge $= \dfrac{\text{purchase price} - \text{residual value}}{\text{estimated life of the asset}}$

Depreciation for a given year
    $=$ the historic cost or NBV (depending on the year)
    $\times$ the percentage expressed as a decimal

Ratio analysis

$$\text{Gross profit margin} = \frac{\text{gross profit}}{\text{revenue}} \times 100$$

$$\text{Operating profit margin} = \frac{\text{operating profit}}{\text{revenue}} \times 100$$

$$\text{Return on capital employed} = \frac{\text{operating profit}}{\text{total equity} + \text{non-current liabilities}} \times 100$$

$$\text{Return on equity} = \frac{\text{operating profit}}{\text{total equity}} \times 100$$

$$\text{Dividend per share (DPS)} = \frac{\text{total dividends}}{\text{number of shares issued}}$$

$$\text{Earnings per share (EPS)} = \frac{\text{net profit after tax}}{\text{number of shares}}$$

$$\text{Dividend yield (DY)} = \frac{\text{dividend per share}}{\text{market price}} \times 100$$

$$\text{Price/earnings ratio} = \frac{\text{market price}}{\text{earnings per share}}$$

Current ratio = current assets ÷ current liabilities

Acid test ratio = (current assets − inventories) ÷ current liabilities

$$\text{Gearing} = \frac{\text{non-current liabilities}}{\text{total equity} + \text{non-current liabilities}} \times 100$$

or

$$\text{Gearing} = \frac{\text{long-term liabilities}}{\text{capital employed}} \times 100$$

Debt to equity ratio = total interest-bearing debt ÷ shareholders' equity

Debt to assets ratio = (total liabilities ÷ total assets) × 100

Proprietary ratio 1 = (shareholders' equity ÷ total assets) × 100

Proprietary ratio 2 = (shareholders' equity ÷ (total assets − intangibles)) × 100

$$\text{Asset turnover} = \frac{\text{revenue (sales)}}{\text{net assets}}$$

$$\text{Inventory turnover} = \frac{\text{cost of sales}}{\text{average inventories held}}$$

$$\text{where average inventories held may be calculated as: } \frac{\text{opening inventories} + \text{closing inventories}}{2}$$

Receivables days = (receivables ÷ sales revenue) × 365

Payables days = (payables ÷ cost of sales) × 365

Investment appraisal

$$\text{Payback} = \frac{\text{income required}}{\text{net cash flow for the year}} \times 12 \text{ (months)}$$

$$\text{Average rate of return} = \frac{\text{net return (profit) per year}}{\text{capital outlay}} \times 100$$

$$\text{Net present value} = \frac{1}{(1 + r)^n} \text{ where } n = \text{numbers of years in the future}$$

Human resources

$$\text{Level of absenteeism} = \frac{\text{number of unauthorised absences}}{\text{total number of days worked by the workforce}} \times 100$$

$$\text{Labour turnover} = \frac{\text{number of leavers that year}}{\text{average number of employees that year}} \times 100$$

$$\text{Labour costs} = \frac{\text{total direct labour costs}}{\text{total units produced}}$$

$$\text{Labour productivity} = \frac{\text{output per time period}}{\text{number of employees}}$$

$Q = \sqrt{2CA \div HP}$ where Q = EOQ; C = acquisition cost per order; A is the total number of units used each month; H is the holding cost as a percentage of the average stock value; and P is the price of each unit

Analysing averages

Variation = actual data − trend data

$$\text{Average seasonal variation} = \frac{\text{total of all seasonal variations for one quarter}}{\text{number of results for this quarter}}$$

Arithmetic mean = sum of items ÷ number of items

Median = $(n + 1) \div 2$ (for an odd number of values) or $n \div 2$ (for an even number of values), where n = the number of values or total frequency

Interquartile range: 1st quartile (Q1) = $1 \times n \div 4$

Mean deviation = $\Sigma(x - \bar{x}) \div n$ where $(x - \bar{x})$ is the difference between each value and the mean; Σ is the mathematical notation for the 'sum of'; n is the number of values. The variance is calculated by $(\Sigma(x - \bar{x})^2) \div n$. The standard deviation is the square root of this.

The mean for normal distribution can be calculated by:

Mean = $n \times p$ where n = the sample size and p = the probability of the event occurring

Remember:

Percentage increase = original $\times$ (1 + the percentage★)

★e.g. 5% increase = 1.05; 15% increase = 1.15

Percentage decrease = original $\times$ (1 − the percentage★★)

★★e.g. 5% decrease = 0.95

Table of discount factors

Below is a table of discount factors for up to 10 years. Most examination boards will provide this information so you only need to learn how to use it.

Year	Discount factors									
	1.00%	2.00%	3.00%	4.00%	5.00%	6.00%	7.00%	8.00%	9.00%	10.00%
0	1	1	1	1	1	1	1	1	1	1
1	0.9901	0.9804	0.9709	0.9615	0.9524	0.9434	0.9346	0.9259	0.9174	0.9091
2	0.9803	0.9612	0.9426	0.9246	0.907	0.89	0.8734	0.8573	0.8417	0.8264
3	0.9706	0.9423	0.9151	0.889	0.8638	0.8396	0.8163	0.7938	0.7722	0.7513
4	0.961	0.9238	0.8885	0.8548	0.8227	0.7921	0.7629	0.735	0.7084	0.683
5	0.9515	0.9057	0.8626	0.8219	0.7835	0.7473	0.713	0.6806	0.6499	0.6209
6	0.942	0.888	0.8375	0.7903	0.7462	0.705	0.6663	0.6302	0.5963	0.5645
7	0.9327	0.8706	0.8131	0.7599	0.7107	0.6651	0.6227	0.5835	0.547	0.5132
8	0.9235	0.8535	0.7894	0.7307	0.6768	0.6274	0.582	0.5403	0.5019	0.4665
9	0.9143	0.8368	0.7664	0.7026	0.6446	0.5919	0.5439	0.5002	0.4604	0.4241
10	0.9053	0.8203	0.7441	0.6756	0.6139	0.5584	0.5083	0.4632	0.4224	0.3855

Notes

Notes